George Henry Hurst

A Dictionary of the Coal Tar Colours

George Henry Hurst

A Dictionary of the Coal Tar Colours

ISBN/EAN: 9783744790727

Printed in Europe, USA, Canada, Australia, Japan

Cover: Foto ©berggeist007 / pixelio.de

More available books at **www.hansebooks.com**

A

DICTIONARY

OF THE

COAL TAR COLOURS.

COMPILED BY

GEORGE H. HURST, F.C.S.

(Member of the Society of Chemical Industry).

LONDON :

HEYWOOD AND CO., LTD., 68, FLEET STREET.

1892.

PREFACE.

THE coal tar colours are now so numerous and the number of synonyms given to them by the various makers is so great that the want of a convenient book of reference dealing with them has been very much felt in this country. English literature is sadly deficient in works relating to the coal tar colours. The only systematic one is that of Benedikt and Knecht, which, although an excellent work, does not give sufficient individual description of each colour.

The writer considered that a work of the nature of a dictionary giving brief descriptions of each coal tar colour would fill the want indicated, hence the appearance of this volume. No systematic account of the process of manufacture, or chemical relationships of the colours, has been attempted, but, as far as it has been possible to obtain the information, the chemical composition, formula, method of making, date of introduction, literature relating to it, and its discoverer have been given, while the properties and uses of the colours have had special attention paid to them.

Besides the descriptions of the coal tar colours themselves, brief accounts of other bodies which are of interest in connection with the subject have been given, and a few special articles added. Every care has been taken to bring the book up to date and to make the descriptions of the colours as accurate as possible, but he is conscious that there must be a few omissions for which, perhaps, readers will pardon him.

His thanks are due to most of the large firms of colour makers for much valuable information regarding their colours which has been embodied in the book.

PLAN OF THE WORK.

In the descriptions of the colours which follow, the name of the colouring matter is given first, then, in brackets, that of the maker, usually in an abbreviated form, which can be readily recognised. The following is a list of coal tar colour makers whose products are noticed :—

Simpson, Maule and Nicholson, now merged in Brooke, Simpson and Spiller, Ltd.
Read Holliday and Sons, Ltd.
Williams, Thomas and Dower, now merged in Williams, Bros. and Co. and Thomas and Dower, Ltd.
Heywood Chemical Co.
Badische Anilin und Soda Fabrik.
Actiengesellschaft für Anilin-Fabrikation (Berlin Aniline Co.).
Farbenfabriken vorm. Fr. Bayer and Co.
Farbwerke vorm. Meister, Lucius and Bruning.
L. Cassella and Co.

A. Leonhardt and Co.
Gesellschaft für Chemische Industrie, Basle (formerly Bindschedler and Busch).
Société Anonyme des Matières Colorantes de St. Denis, Paris (formerly Poirrier and Co.).
Durand and Huguenin.
Gilliard, P. Monnet and Cartier.
R. Geigy and Co.
Kalle and Co.
Dahl and Co.
K. Oehler.
Wm. Noetzel and Co.

Following the name of the maker, is the chemical name, formula and method of making. Then follows the date, in brackets, of its introduction, with the name of the discoverer. Reference is given also to the patent and to other literature relating to he colouring matter. The rest needs no explanation.

INTRODUCTION.

——:0:——

SINCE the year 1856, when Perkins first introduced his Mauve to the world, a very large number of coal tar colouring matters suitable for the dyeing and printing of textile fibres have been discovered by the indefatigable researches of chemists. These are derived from the many compounds which can be extracted from coal tar, and from this fact are commonly spoken of as the coal tar colours. The number of colours known to the chemist is very great. Some of these, which will be found described in the following pages, have been, and are in use to an enormous extent in the textile colouring trades, in which industries they have to a very large extent displaced the natural dyestuffs, since in their brilliance of tint, ease of application and fastness, they present many advantages over them. Many colouring matters derived from coal tar have not come into use owing to a variety of circumstances. Want of brilliance of colour, difficulty of application, or lack of permanence are faults common to many of them, and this last is a very great drawback. It has caused the rejection of many colouring matters which have been produced in the laboratory, and the withdrawal of not a few which were placed on the market in the early days of the colour industry.

The coal tar colours may be divided into 16 groups :—

1. *Nitro Colouring Matters.*—Picric Acid, Naphthol Yellow, &c. r. NITRO COLOURING MATTERS.

2. *Nitroso Colouring Matters.*—Resorcin Green, Gambine, &c. v. NITROSO COLOURING MATTERS.

3. *Azoxy Colouring Matters.*—There are few dyestuffs at present known belonging to this group, the characteristic feature of which is their containing the group —N—O—N—, the nitrogen atoms being capable of acting as a binding group to two radicles. Curcumine S is an example.

4. *Azo Colouring Matters.*—Aniline Yellow, Crocein Scarlet, Congo Red. v. AZO COLOURING MATTERS.

5. *Hydrazin Colouring Matters.*—These contain the group
$$\begin{matrix} -CN-NH- \\ | \\ -CN-NH- \end{matrix}$$
They are but few in number. Phenanthrene Red is the most familiar member of the group.

6. *Diphenyl Methane Colouring Matters.*—The few dyestuffs belonging to this group are derivatives of diphenyl methane, and contain the group
$$C \begin{cases} C_6H_4 \\ =NH \\ C_6H_4 \end{cases}$$
Auramine is the commonest member of the group. They are essentially basic dyestuffs, and are in the main fast and therefore very useful.

7. *Rosaniline Colouring Matters.*—Magenta, Violets, Blues, Brilliant Green. v. ROSANILINE COLOURING MATTERS.

8. *Anthracene Colouring Matters.*—Alizarine, Alizarine Orange, &c. v. ALIZARINE COLOURING MATTERS.

9. *Indophenols.*—r. INDOPHENOL.

10. *Orazine Colouring Matters.*—Meldola's Blue, or New Blue, is almost the solitary representative of this class of dyestuff, which contains the group $\left\langle \begin{matrix} N \\ O \end{matrix} \right\rangle$ connected with a diad radicle on each side. v. MELDOLA'S BLUE.

11. *Thionine Colouring Matters.*—Containing sulphur in combination with two organic radicles. r. LAUTH'S VIOLET.

12. *Eurhodine.*— r. EURHODINE.

13. *Azine Colouring Matters.*—Safranine, Magdala Red.   r. AZINES.

14. *Induline and Nigrosine Colouring Matters.* —r. INDULINE.

15. *Chinoline and Acridine Colouring Matters.*—r. ACRIDINE ORANGE and QUINOLINE YELLOW.

16. *Artificial Indigo.*—Several methods of preparing indigo from compounds derived from coal tar products have lately been announced, but as yet, owing to a variety of circumstances, these methods have not come into practical use. The consideration of artificial indigo has not been attempted in this work.

Most of these groups will be found described in special articles in the body of the book. Even this classification cannot be considered as a permanent one, and there is no doubt that as time goes on and new dyestuffs are discovered a new rearrangement into groups will be requisite.

The question of the chemical constitution of the coal tar colours is a very interesting one, especially in so far as it relates to the cause of their colour and in so far as it points out the influence the various elements, or the manner in which they are grouped together, has on the colour and properties of the dyestuffs.

From the manner in which the colouring matters comport themselves when caused to undergo chemical action, and also from the methods of preparation of them, chemists have been able to recognise in them specific groups of elements or atoms, which in many cases impart to the colour certain characteristic properties. Thus many compounds contain a group of three atoms, one of nitrogen, two of hydrogen, known as amidogen and having the formula NH_2. This group of atoms, whenever present, imparts basic properties. Again, the group OH gives, when present, what is termed a phenolic character.

The groups of atoms which enter into the composition of coal tar colours may be divided into two divisions :—1st. Those which have an influence on, or are a cause of colour in a compound. 2nd. Those groups to which the acid or basic character of the colouring matter is due.

The first division of these groups of atoms is called *Chromophores*, meaning colour-bearing groups. Thus the group NO_2, nitryl, is present in a large number of colouring matters. Picric acid is trinitrophenol $C_6H_2(NO_2)_3OH_3$, and that the yellow-colouring power is due to the NO_2 groups is quite certain from the fact that if picric acid be acted upon by nascent hydrogen it loses the oxygen of these groups and is converted into triamidophenol $C_6H_2(NH_2)_3OH$, which has no colouring power whatever. Now the colouring matters which, like picric acid, contain this group NO_2, nitryl, are characterised by being yellow dyestuffs capable of dyeing animal fibres, wool, silk, etc., in acid baths.

The group NO, nitrosyl, is also a chromophore, but the dyestuffs containing it are only capable of forming a distinctive colour when combined with a metallic compound; in other words, they are mordant dyeing dyestuffs. Gambine and Resorcin Green are examples of this group.

The azo colours contain the chromophorous group —N = N— which binds together two organic radicles, thus: $C_{10}H_7$—N = N—$C_{10}H_6OH$, naphthylene-azo-naphthol, a scarlet dye. Although generally speaking the group —N = N— is a colour-bearing one, yet it does not follow that all compounds containing it are colouring matters. Much depends upon the nature of the radicles with which it is combined ; for instance, diazo-benzene chloride C_6H_5—N = N—Cl is not a colouring matter although it contains this group. The bodies into which these chromophores enter are called chromogenes and comprise such as nitrobenzene $C_6H_5NO_2$; azo-benzene C_6H_5—N = N—C_6H_5 ; diphenylimide

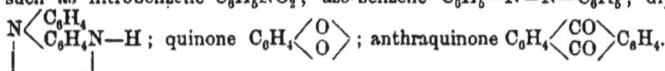

$$N\begin{matrix} C_6H_4 \\ C_6H_4 \end{matrix}N\!-\!H \; ; \text{ quinone } C_6H_4\!\!\left\langle\begin{matrix}O\\O\end{matrix}\right\rangle ; \text{ anthraquinone } C_6H_4\!\!\left\langle\begin{matrix}CO\\CO\end{matrix}\right\rangle C_6H_4.$$

These are generally colourless, or at the most slightly coloured, possessing little or no dyeing power, but by the introduction of chromophorous groups they become dyestuffs. It is of interest to note how the internal arrangement of the atoms of a compound influences its dyeing properties. Thus there are four bodies known which have the formula $C_{19}H_{18}N_3Cl$— two of these are colourless, one is yellow and the other is green when solid, red in solution. These bodies are the hydrochlorides of triphenyl-guanidin, the hydrochloride of the amidine of *p*-amido-benzoic acid, the hydrochloride of diamido-benzophenonephenylimide (an auramine base) and the hydrochloride of pararosaniline (Magenta). So far as the constitution of these bodies are known they have the atomic arrangement shown in the formulæ

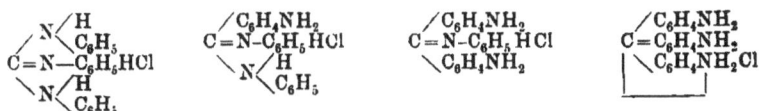

$$C=N-C_6H_5\underset{\displaystyle N<\frac{H}{C_6H_5}}{\overset{\displaystyle <\frac{N<\frac{H}{C_6H_5}}{}}{}}HCl \qquad C=N-C_6H_5 \cdot HCl\underset{\displaystyle N<\frac{H}{C_6H_5}}{\overset{\displaystyle <\frac{C_6H_4NH_2}{}}{}} \qquad C=N-C_6H_5\underset{\displaystyle C_6H_4NH_2}{\overset{\displaystyle <\frac{C_6H_4NH_2}{}}{}}HCl \qquad C=C_6H_4\underset{\displaystyle C_6H_4NH_2Cl}{\overset{\displaystyle <\frac{C_6H_4NH_2}{}}{}}$$

The groups OH and NH_2, when they enter into combination with chromogenes, impart to them dyeing properties, OH (hydroxyl) giving acid or mordant dyeing properties, while NH_2 (amidogen) converts the chromogen into a basic dye.

These groups have been named auxochromes, and upon their number and arrangement depends a great deal the colouring power of the dyestuff. Thus there are three amido-azo-benzenes.

Amido-azo-benzene (Aniline Yellow), $C_6H_4NH_2—N=N—C_6H_5$.

Diamido-azo-benzene (Chrysoidine), $C_6H_4NH_2—N=N—C_6H_4NH_2$.

Triamido-azo-benzene (Bismarck Brown), $C_6H_3(NH_2)_2—N=N—C_6H_4NH_2$.

As the number of NH_2 groups increases so does the colour and dyeing power. In the same way the introduction of more NO_2 groups increases the dyeing power; thus mononitrophenol $C_6H_4NO_2OH$ has little or no colour, while trinitrophenol $C_6H_2(NO_2)_3OH$ is a strong yellow dye. Again, alizarine is dioxyanthraquinone $C_6H_4<^{CO}_{CO}>C_6H_2(OH)_2$, and gives a scarlet with alumina; Purpurin is trioxyanthraquinone $C_6H_3OH<^{CO}_{CO}>C_6H_2(OH)_2$, and gives a crimson with alumina; Alizarine Bordeaux is tetroxyanthraquinone $C_6H_3(OH)_2<^{CO}_{CO}>C_6H_2(OH)_2$, and gives a dark red with alumina; while Alizarine Cyanine is pentoxyanthraquinone $C_6H(OH)_3<^{CO}_{CO}>C_6H_2(OH)_2$, and gives a red-violet with alumina. The colouring power depends upon the fact that the OH and NH_2 groups impart to the body salt-forming properties. If these be destroyed (as they can be) then the colouring power is destroyed—thus, acetyl-amido-azo-benzene $C_6H_5N=NC_6H_4N<^{H}_{C_2H_3O}$ has but little colour.

The acid or basic character of the dyestuffs is determined by the presence or otherwise of certain groups which cause them to be basic, or acid, or neutral. As it has been previously mentioned, the auxochrome group NH_2 imparts basic properties to those colouring matters containing it, and the auxochrome group NO_2 imparts acid properties to all colouring matters containing it. SO_3H, the sulphon group, while not influencing the character of the colour, imparts acid properties. The auxochrome group OH imparts weak acid properties, and its salts are readily decomposed by acids. Even a weak acid such as acetic is capable of effecting this change. The acidic character of colouring matters containing hydroxyl is much affected by the presence of other bodies, such as Cl, I, Br, NO_2, which generally increase the acid property. Thus picric acid $C_6H_2(NO_2)_3OH$ is a stronger acid than phenol C_6H_5OH; Eosin, tetrabromfluorescein $C_{20}H_6Br_4O_5(OH)_2$ is a much stronger acid than fluorescein $C_{20}H_{10}O_5(OH)_2$. The solubility of the colouring matters containing hydroxyl is but slight: alizarine has a solubility of one part in 3,000 parts of water; picric acid one in 86; eosin is insoluble; fluorescein will dissolve only in boiling water. The articles on ROSANILINE, AZO COLOURING MATTERS, NITRO COLOURING MATTERS, and ALIZARINE may also be consulted on this subject.

DICTIONARY

OF THE

COAL TAR COLOURS.

A

Acid Black B—

(READ HOLLIDAY AND SONS.)—A disazo colouring matter. (1890).—*Dark crystalline bronzy powder*, soluble in water to a violet solution, very slightly soluble in alcohol, soluble in acetic acid to a violet solution, in strong sulphuric acid to an olive green solution, on diluting with water a violet precipitate is obtained. Hydrochloric acid has no action when added to an aqueous solution. Caustic soda has no action.—*Dyes* wool and silk from acid baths, red violet blacks, fast to acids, light, but not to soaping.

Acid Black BB—

(READ HOLLIDAY AND SONS.)—A disazo colouring matter. (1890).—*Dark bronzy blue powder*, soluble in water to a dark violet blue solution, slightly soluble in alcohol to a greenish blue solution, in acetic acid to a violet blue solution, in strong sulphuric acid to a dirty green solution, which on diluting with water turns violet blue. Hydrochloric acid added to the aqueous solution turns it bluer. Caustic soda has no action.—*Dyes* wool and silk from acid baths, violet blacks, fast to acid, light and alkalies, not to soaping.

Acetin Blue—

An indulin blue insoluble in water ; introduced by the Badische Anilin und Soda Fabrik for printing when dissolved in acetin (1886).—See Indulin.

Acid Brown G—

(BER. ANIL.) — Benzene-azo-*m*-diamido benzene azo-benzene-*p*-sodium sulphonate

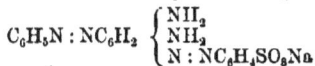

$$C_6H_5N : NC_6H_2 \begin{cases} NH_2 \\ NH_2 \\ N : NC_6H_4SO_3Na \end{cases}$$

obtained by combining aniline with *meta*-diamido - azo - benzol - *p* - monosulphonic acid. Discovered in 1882.—*Brown powder*, soluble in water with a brown colour, and in strong sulphuric acid to a red-brown solution ; on diluting with water a yellow brown solution is obtained. Applied to wool in an acid bath.

Acid Brown R—

(BER. ANIL.)—Sodium naphthionate azochrysoidine

$$C_{10}H_6 \begin{cases} SO_3Na \\ N : N \end{cases} \Big\} \ C_6H_3N : N \Big\} \ C_6H_3 \begin{cases} NH_2 \\ NH_2 \end{cases}$$

obtained by combining sodium naphthionate with chrysoidine. Discovered in 1882. —*Brown powder*, dissolves in water to a brown coloured solution, in strong sulphuric acid to a dirty, olive-coloured solution, on diluting with water first turns reddish, then gives a brown precipitate.— —*Dyes* wool brown in acid bath.

Acid Cerise—

An impure form of acid magenta.

Acid Green—

Under this term greens derived from two different bases come into the market.

(1.) The sodium salt of tetramethyl-diamidotriphenylcarbinol monosulphonic acid.

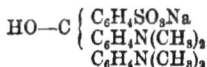

$$HO—C \begin{cases} C_6H_4SO_3Na \\ C_6H_4N(CH_3)_2 \\ C_6H_4N(CH_3)_2 \end{cases}$$

obtained by sulphonating benzaldehyde green or by sulphonating tetramethyl-diamidotriphenylmethane and oxidizing the product. The first method was discovered in 1878 and the second in 1880.— *Bright green powder*, slightly soluble in water with a blue-green colour. On adding hydrochloric acid red brown crystals of the sulphonic acid form ; on adding caustic soda to the aqueous solution it remains clear, but is decolourised. In strong sulphuric acid it is slightly soluble to a light yellow solution, on diluting with water it is first reddish yellow, gradually changing to yellow green, and lastly to green.—*Dyes* wool and silk green in acid baths, and also mordanted cotton. Comes also into the market as Helvetia green,

(2.) The sodium salt of drethyl or dimethyl dibenzydiamidotriphenylcarbinol trisulphonic acid.

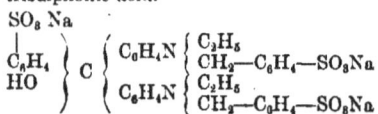

$$SO_3Na$$

$$C_6H_4 \atop HO \Big\} C \Big\{ {C_6H_4N \Big\{ {C_2H_5 \atop CH_2-C_6H_4-SO_3Na} \atop C_6H_4N \Big\{ {C_2H_5 \atop CH_2-C_6H_4-SO_3Na}}$$

obtained by condensing benzaldehyde with benzylethylanilin, sulphonating the product and oxidizing the sulphonic acid formed. Discovered in 1879. — Bright green powder, soluble in water to green solution, soluble also in alcohol. Hydrochloric acid changes the colour of the aqueous solution to a yellow brown. Caustic soda discolours and produces a dirty violet turbid appearance. Soluble in strong sulphuric acid with a yellow colour; on adding water gradually turns green.—*Dyes* silk and wool in acid baths. Comes into the market as light green S, light green SF, acid green SOF.

Acid Magenta—

(READ HOLLIDAY AND SONS.)—Usually the sodium salt of the trisulphonic acid of rosaniline.

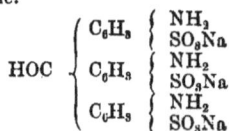

$$HOC \Big\{ {C_6H_3 \Big\{ {NH_2 \atop SO_3Na} \atop C_6H_3 \Big\{ {NH_2 \atop SO_3Na} \atop C_6H_3 \Big\{ {NH_2 \atop SO_3Na}}$$

obtained by treating magenta with fuming sulphuric acid. Discovered in 1875. Occurs in the form of green grains or powder with a metallic lustre; soluble in water with a blue-red colour, but nearly insoluble in alcohol; caustic soda added to an aqueous solution causes almost complete discolourisation. Soluble in strong sulphuric acid to a yellow solution, on dilution with water the colour is gradually changed to red.—*Dyes* wool and silk in acid baths, but is not suitable for cotton. Comes into the market as fuchsin S, rubin S, acid rubin, acid maroon, acid claret, etc., in more or less pure forms.

Acid Mauve B—

(READ HOLLIDAY AND SONS.)—A sulphonate of pure mauvaniline.—*Bronzed colour-ed* lustrous crystals soluble in water to a dark violet red solution, in strong sulphuric acid to a dark red solution; on diluting with water a dark brownish yellow solution is obtained. Caustic soda turns the colour of the aqueous solution more scarlet.—*Dyes* wool and silk from acid baths fine shades of mauve, moderately fast to acids, washing, not fast to light.

Acid Ponceau—

(DURAND AND HUGUENIN.)—Sodium salt of β-naphthylamine sulphonic acid-azo-β-naphthol.

$$C_{10}H_6SO_3NaN : NC_{10}H_6OH$$

obtained by diazotising β-naphthylamine monosulphonic acid and combining with β-naphthol.—*Scarlet red powder*, only slightly soluble in cold water, but easily soluble in warm water to a scarlet solution; in strong sulphuric acid to a crimson solution, on diluting with water a brown precipitate falls down. Addition of hydrochloric acid to the aqueous solution throws down a brown precipitate. Caustic soda turns the colour of the aqueous solution brown.—*Dyes* wool in acid baths red. This colour is isomeric with fast brown 8 B, double brilliant scarlet G, fast brown (Badische), fast red A.

Acid Roseln—

Syn. of acid magenta.

Acid Rubine—

Syn. of acid magenta.

Acid Maroon—

An impure form of acid magenta.

Acid Violet 6B—

(BAYER.)—The pentamethylbenzyl para-rosaniline sodium sulphonate.

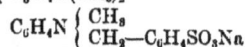

$$HO-C \Big\{ {C_6H_4N(CH_3)_2 \atop C_6H_4N(CH_3)_2}$$

$$C_6H_4N \Big\{ {CH_3 \atop CH_2-C_6H_4SO_3Na}$$

obtained by the oxidation of pentamethyl-benzylparaleukanilin monosulphonic acid. Discovered in 1883.—*Dark violet powder*, soluble in water with a violet colour, in alcohol with a bluer colour. Hydrochloric acid added to the aqueous solution turns it first green, then yellow. Soluble in concentrated sulphuric acid to a brownish yellow solution, on diluting with water turns first green then blue, lastly violet. *Application.*—Dyes wool and silk in acid bath a bright violet.

Acid Violet 6B—

(BADISCHE.)—See Acid Violet (*Bayer*).

Acid Violet N—

(MEISTER LUCIUS.) —1890— *Dark reddish violet powder*, soluble in water to a violet solution, in strong sulphuric acid to an amber coloured solution, which becomes paler on diluting with water. Hydrochloric acid added to the aqueous solution turns it olive yellow, caustic soda has no action.—*Dyes* wool and silk from acid baths bright bluish violet, dilute acids have no action, strong acids turn it dull red, caustic soda no action. Not fast to soaping.

Acid Violet 4RS—

(BADISCHE.)—Sodium salt of dimethyl rosaniline trisulphonic acid.

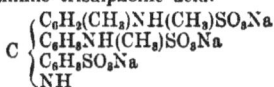

$$C \begin{cases} C_6H_2(CH_3)NH(CH_3)SO_3Na \\ C_6H_4NH(CH_3)SO_3Na \\ C_6H_3SO_3Na \\ NH \end{cases}$$

obtained by treating dimethyl rosaniline with fuming sulphuric acid. Discovered in 1877. *Eng. Pat.* No. 3731, 1877.— *Red violet powder*, slightly soluble in water with a magenta colour; on adding caustic soda and warming, the colour changes to a feeble reddish yellow. Soluble in strong sulphuric acid to a brownish yellow solution, on diluting with water the colour is restored.—*Dyes* wool and silk in acid bath a bluer shade than acid magenta. Known also as red violet 4RS.

Acid Yellow—

Under this name two different products are in the market and used by dyers.

1.—*Acid Yellow.* (WILLIAMS, THOMAS AND DOWER.)—Sodium salt of sulphanilic acid-azo-diphenylamine

$$C_6H_4 \begin{cases} SO_3Na \\ N : NC_6H_4—CH—C_6H_5 \end{cases}$$

obtained by combining sulphanilic acid with diphenylamine. Discovered in 1876 by Witt. Roussin and Poirrier. *Eng. Pat.* 4491, 1878. — *Orange yellow plates*, soluble in water to orange yellow solution, to which the addition of hydrochloric acid causes the formation of a violet precipitate. Soluble in sulphuric acid to violet solution, on addition of water a violet precipitate is thrown down.—*Dyes* wool in an acid bath, an orange yellow colour. This colour is also known as diphenylamine orange, diphenyl orange, orange IV, *Badische*, Tropœolin OO (original name) orange M, (Socy. Chem. Ind. Basle) orange G. S., new yellow *(Bayer)*.

2.—*Acid Yellow* (Actiengesellschaft). A mixture of sodium amidoazobenzol disulphonate with the monosulphonate.

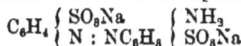

$$C_6H_4 \begin{cases} SO_3Na \\ N : NC_6H_3 \end{cases} \begin{cases} NH_3 \\ SO_3Na \end{cases}$$

obtained by the action of fuming sulphuric acid on the hydrochloride of amidoazobenzol. Discovered in 1878.—*Yellow powder*, soluble in water to a yellow solution, changing to orange on addition of hydrochloric acid. Soluble in concentrated sulphuric acid to a brown yellow solution, changing to orange yellow on dilution with water.—*Dyes* wool and silk in acid baths a yellow. This colour is also known as fast yellow G, acid yellow G, fast yellow extra, F fast yellow, new yellow, solid yellow.

Acid Yellow—

Syn. of naphthol yellow S.

Acid Yellow D—

Syn. of acid yellow (1).

Acid Yellow G—

Syn. of acid yellow (2).

Acid Yellow S—

Syn. of naphthol yellow S.

Acridine Orange—

(LEONHARDT.)

$$\left[N(CH_3)_2C_6H_3 \left\langle \begin{matrix} CH \\ N \end{matrix} \right\rangle C_6H_3N(CH_3)_2HCl \right]_2 ZnCl_2$$

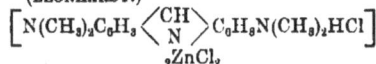

1890—English patent, 3243, 1890.—*Pale brownish red powder*, soluble in water to an orange solution, in alcohol to a yellow solution with a yellow green fluorescence, in acetic acid to an amber coloured solution with a yellowish green fluorescence, in strong sulphuric acid to a pale yellow solution with a green fluorescence; on diluting with water a reddish orange solution is obtained. Caustic soda added to the aqueous solution gives a bright yellow precipitate in a pale yellow solution which is fluorescent.—*Dyes* cotton mordanted with tannic acid and tartar emetic, wool and silk in neutral baths, dark reddish orange shades; acids redden the shade a little; does not stand soaping; fairly fast to light.

Aldehyde Green—

A derivative of trichinolymethane $C_{22} H_{27}N_3S_2O$ obtained by treating rosaniline sulphate with aldehyde until the product is soluble in water with a green colour, then pouring into a boiling solution of thiosulphate of soda, boiling and filtering. The colouring matter forms a green, amorphous powder; insoluble in water. very slightly in alcohol, but soluble in a mixture of alcohol, water and sulphuric acid. Discovered in 1861 by Cherpin, French patent, 26 June, 1861.—*Dyes* wool silk green in acid bath. Not now used. Known as usebe green and aniline green.

Alizarine—

In 1826, Colin and Robiquet showed that the root of the madder plant *(Rubia tinctoria)* contained as its chief colouring principle what they called *alizarine*, from the Levantine name of the plant, *Alizari*. Schunck subsequently confirmed their researches, and gave to alizarine the formula $C_{14}H_{10}O_4$. From the madder root was also obtained purpurin, another colouring principle. Graebe and Liebermann, about 1866, succeeded in obtaining an hydrocarbon body having the formula $C_{14}H_{10}$ from alizarine prepared from the madder, and this they proved to be identical with the anthracene previously obtained by Dumas and Laurent from coal tar, in which it exists in large quantities. Graebe and Liebermann had already ascertained that alizarine was a dihydroxyquinone, and so concluded that it must be derived from anthracene and related to it, as shown by the formulæ—

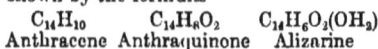

$$C_{14}H_{10} \quad C_{14}H_8O_2 \quad C_{14}H_6O_2(OH_2)$$
Anthracene Anthraquinone Alizarine

In 1862 Anderson had, in the course of his investigations, discovered anthraquinone, which he had named oxyanthracene. The problem now became simply to find out a process to convert anthracene into alizarine. Graebe got one clue to this in the course of his previous researches on the quinones; he took anthraquinone and treated it with bromine and converted it into dibromanthraquinone, which on fusion with caustic potash yielded alizarine and potassium bromide. This they patented in December, 1868. This process is, however, too expensive to be a commercial success.

Perkin had been working with a view to find a better process, and such a one he patented on the 26th June, 1869, which consisted in treating anthraquinone with sulphuric acid, whereby it is converted into disulphonic acid, which when fused with potash gave alizarine. Graebe, Caro, and Liebermann patented the same process on 25th June, 1869. The Badische worked this latter patent.

Dr. Perkin made alizarine commercially, afterwards he sold his patent rights to Messrs. Brooke, Simpson and Spiller, who resold them to Messrs. Burt, Boulton and Haywood, who again on the expiration of the patents sold them to the British Alizarine Company.

Anthracene has the formula

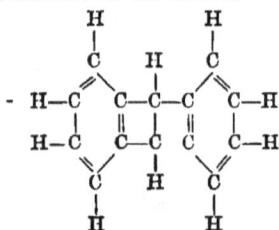

and contains two benzene residues; on oxidation it is converted into anthraquinone

$$C_6H_4\Big\langle{CO \atop CO}\Big\rangle C_6H_4$$

This body can be converted into dihydroxyanthraquinones, of which there are

eight isomers known, of which alizarine is one.

These isomers are due to the fact that the hydroxyl groups may be situated in the same benzene nucleus or one in each benzene nucleus. Some of these isomeric dioxyanthraquinones are of commercial interest: these will be described; others are not and will only be named.

1.—*Alizarine.*

$$C_6H_4\Big\langle{CO \atop CO}\Big\rangle C_6H_2\Big\langle{OH_1 \atop OH_2}$$

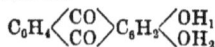

This occurs in the root of the madder plant (*Rubia tinctoria*); partly free, partly in the form of a glucoside, rubery-thric acid which under the influence of fermentation yields alizarine and glucose.

Alizarine is obtained from anthracene by first converting it into anthraquinone, which is done by treating it with a mixture of bichromate of potash and sulphuric acid. This anthraquinone, after purification, is converted into the disulphonic acid by treatment with fuming sulphuric acid; this is then fused with caustic potash, when alizarine is formed. The crude alizarine melt contains various impurities — monohydroxyanthraquinone, anthrapurpurin, flavopurpurin, etc.; these are separated out by purifying processes.

Alizarine can be obtained from alcohol in the form of red needles melting at 282deg. C., subliming at higher temperature to magnificent deep red prisms. Dissolves in alkaline solutions with a deep purple colour, alkaline alizarates being formed. These alkaline solutions are precipitated by salts of the alkaline earths and earthy metals, insoluble coloured alizarates being formed. Alizarine therefore re-acts like a weak acid. Soluble in strong sulphuric acid to a red solution. The commercial product is sold in the form of an orange brown paste containing 20 per cent. of actual alizarine.

2.—*Quinizarine* is the 1 : 4 dihydroxyanthraquinone

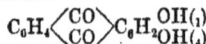

$$C_6H_4\Big\langle{CO \atop CO}\Big\rangle C_6H_2{OH_{(1)} \atop OH_{(4)}}$$

This body is of no practical importance.

3.—*Anthraflavic acid,*

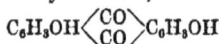

$$C_6H_3OH\Big\langle{CO \atop CO}\Big\rangle C_6H_3OH$$

is produced by fusing a disulphonic acid of anthraquinone with potash. It is present in commercial alizarine. It differs from alizarine in forming yellow needles melting at 330deg. C., in dissolving in alkalies to a yellowish red solution, in strong sulphuric acid to green solution. Anthraflavic acid has no dyeing properties.

4.—*Isoanthraflavic acid* is formed from

B - anthraquinone disulphonic acid by fusion with potash, and is found in the crude alizarine melt. Forms yellow needles, soluble in alkalies to a red solution ; it does not dye cloth. Fused with potash it gives anthrapurpurin.

5.—*Anthrarufin,*

$$(_4)OHC_6H_3\!\!<\!\!^{CO}_{CO}\!\!>\!\!C_6H_3OH(_1)$$

forms yellow needles, melting at 280deg. C., soluble in strong sulphuric acid with a deep red colour só intense that one part of this solution is distinguishable when mixed with ten million parts of water.

6.—*Chrysazin* 1 : 3 dihydroxyanthraquinone is of little interest.

Alizarine is the only one of these that has commercial interest. It dyes fibres which have been previously mordanted. Alumina gives bluish shades of fiery red ; chrome gives darker shades of red ; iron gives dark blue violet shades. Alizarine comes into commerce in the form of a brown-yellow paste containing, as a rule, 20 per cent. of actual colouring matter, and all the red anthracene colours are known as alizarine, various marks. Alizarine blue shade, or alizarine yellow shade.

Alizarine V2 (BADISCHE), alizarine Nr. 1 (MEISTER), alizarine 1e (BAYER), alizarine 1a (LEVERKUS) is alizarine.

Alizarine G1 (BADISCHE), alizarine SOG (MEISTER), alizarine X (BAYER), alizarine FA (LEVERKUS) is flavopurpurin.

Alizarine GD (BADISCHE), alizarine RX (MEISTER), alizarine SX (BAYER), alizarine RF (LEVERKUS) is anthrapurpin.

Alizarine No. 6 (MEISTER) is purpurin.—See also anthrapurpurin, purpurin, flavopurpurin, anthragallol.

Literature.—*Perkin. Chem. Socy. Jour.* 1870, p. 133 ; 1871, p. 1109 ; 1872, p. 19. Levinstein. *Jour. Socy. Chem. Ind.* 1883, p. 213. *Chemical News,* 1870. Perkin. *Jour. Soc. of Arts,* 1879.⸴

Alizarine Maroon—

(BADISCHE.)—Bisulphite of sodium combination of a product formed by the action of sulphuric acid on alizarine blue. (1889.) *Eng. Pat.* 14353, 1888. — *Chocolate coloured paste,* slightly soluble in cold, easily in hot water to a dull red solution. Hydrochloric acid gives a dirty red precipitate, strong sulphuric acid dissolves it with evolution of nitrous fumes to an amber coloured solution, on diluting with water a flocculent precipitate in an orange coloured solution is obtained. Caustic soda gives a purple solution.—*Dyes* wool mordanted with chrome, dull reddish purples, slightly altered by acids and alkalies ; fast to soaping.

Alizarine N—

(LEVERKUS.)—Syn. of alizarine orange.

Alizarine WS I—

(LEVERKUS.)—Syn. of alizarine.

Alizarine Colours in Dyeing and Printing—

To describe in detail the various methods and processes by means of which alizarine and its allied colouring matters, alizarine orange, purpurin, anthrapurpurin, anthragallol, gallein, alizarine blue, alizarine yellow, &c., are applied in dyeing and calico printing, would take too much space for this work ; it is therefore proposed to give simply an outline of the principal methods at present in use.

COTTON DYEING.

All the alizarine colours are not suitable for this purpose. The cotton requires to be mordanted. The mordant most commonly used is alumina in one or other of its forms. Iron is occasionally used. Acetate of lime is also used, and generally the cotton is prepared with Turkey red oil or oleine before mordanting. The oil has a brightening action on the finished colour.

The yarn or cloth which has been bleached is thoroughly wet out, and then padded in a cold solution of oleine or Turkey red oil, 1 of 50 per cent. oleine in 10 of water, wrung out and stove-dried. This process is repeated. When dried it is then steamed for 1 hour at 2lb. pressure. Next, the cotton is worked well for half an hour, and then allowed to steep for 2 to 4 hours in acetate of alumina 9deg. Twaddell. If this contains 1oz. of tin crystals, the colour will be brighter. Instead of acetate of alumina, the basic sulphate may be used. The goods are wrung out, run through a chalk bath, containing 2oz. of chalk per gallon. The goods are now ready for dyeing. This operation is carried out as follows :—The dye-bath is made with 10 to 15 per cent. of alizarine with 1 per cent. of chalk. The cotton is worked in the cold for half an hour, then the temperature is raised to the boil, and continued at this temperature until the bath is exhausted. The goods are now wrung out, steeped in a new oil bath as before, then steamed for one hour and cleared by boiling in soap. Turkey-red dyeing is done very much on the same principles, only the operations are repeated several times. Any other mordant may be substituted for the alumina. For dyeing all the so-called alizarine colours on cotton the following process may be used with fairly good results :— 1st. The goods are prepared in the usual way with Turkey-red oleine ; they are then dried and steamed. 2nd. The goods are then entered into a dye-bath, containing 5 to 20 per cent. of the colouring matter, and 2 per cent. of acetic acid. In this bath the goods are worked for half an hour in the cold, then slowly heated to from 180deg. to 190deg. F., and maintained at that heat until the dye-bath is exhausted. The cotton is then wrung out. 3rd. The dyed goods are now entered into a bath of 1gal. chromium acetate 30deg. Twaddell, in 10gal. of water at about 180deg. F; the temperature is raised to the boil and kept gently boiling for one hour; the goods are now well rinsed in cold water and soaped well.

CALICO PRINTING.

There are two methods of applying alizarine colours in calico printing. 1st. By direct printing. 2nd. By printing and dyeing. *1st. Direct Printing.*—A printing colour is made with—

> 6qts. water
> 1 ,, acetic acid 8deg. Tw.
> ½ ,, sulphocyanide of alumina 20deg. Tw.
> ¾ ,, alizarine 20 per cent.
> ¼ ,, glycerine
> 2¾lb. starch

Boil, cool, print, steam for 1½ hour at 3lb. pressure, soap and dry. By varying the quantity and quality of the alizarine various shades of red can be got, by using iron instead of alumina violets are got, chocolates by mixing the two mordants. For those colours like alizarine blue, galloflavin, anthracene brown, which require a chrome mordant, the following printing colour may be used as a standard:

> 7¼lb starch
> 1¼ ,, dextrine
> ½gal. acetic acid 9deg. Tw.
> 1pt. olive oil
> 3gal. water

Boil, allow to cool, and add—

> 25lb. colouring matter
> ½gal. acetate of lime 24deg. Tw.
> 7pts. acetate of chrome 33deg. Tw.
> 6pts. acetic acid 9deg. Tw.

Prepare the cloth with oleine, print' steam for one hour, pass through chalk bath, and soap. This will usually give dark shades, by adding thickening lighter shades can be got. For the acetate of chrome in the above standard, acetate of alumina or acetate of iron may be substituted if required.

WOOL DYEING.

1st.—Chrome mordant.—The wool is first mordanted by boiling for 1½ hour with 3 per cent. of bichromate of potash and 1 per cent. of sulphuric acid, rinsed

and then entered into the dye-bath which contains the colouring matter, plus a small quantity of acetic acid. The wool is entered and worked for half an hour in the cold; then the bath is gradually heated to the boil, and kept boiling for 1¼ to 2 hours. The goods are then rinsed and dried.

2nd.—Alumina mordants.—A mordanting bath is made with 5 to 8 per cent. of sulphate of alumina with 3 to 5 per cent. of tartar. The wool is worked cold for from 15 to 20 minutes, then raised to the boil and boiled for 15 to 30 minutes; after washing, the dyeing is done as with chrome mordants.

3rd.—Iron mordants.—As with alumina, using 3 to 5 per cent. ferrous sulphate and 2 to 3 per cent. tartar.

SILK DYEING.

1st.—Alumina mordants.—The mordanting bath is made with 4½lb. alum, 7½oz. soda crystals, and 7½gal. of water. Boil the bath until it becomes clear, when cold enter the silk, work well for half an hour, then allow to steep for 12 hours. Wring out. The mordanting bath is not exhausted, and can be used again after adding fresh material. The dye-bath is made with boiled off liquor, which must be as neutral as possible, and the colouring matter. Work the silk in it cold for 20 minutes, then heat to boil and dye at the boil for one hour; rinse, soap slightly, and brighten in acetic acid.

See Hummel. Alizarine colours in wool dyeing. *Jour. Socy. Dyers and Colourists.* Vol. I., 1885, p. 11.

Sansone. Alizarine dyeing and printing on cotton. *Jour. Socy. Dyers and Colourists.* Vol. I., 1885, p. 203.

Alizarine Black S—

(BADISCHE.)—Sodium bisulphite compound of naphthazarine.

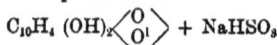

$$C_{10}H_4 (OH)_x \!\!\begin{array}{c} O \\ O' \end{array}\!\! + NaHSO_3$$

Naphthazarine or dioxynaphthaquinone is formed by treating dinitronaphthalene with zinc and strong sulphuric acid, and this then acted on by sodium bisulphite yields alizarine black S. Naphthazarine has been known since 1861, when it was discovered by Roussin. Its bisulphite compound was only introduced in 1887. English patent 7833, 1877.—*Black paste*, insoluble in cold, but easily soluble in hot water to a red-brown solution; soluble in alcohol to a yellow-brown solution, having a green fluorescence. Soluble in cold strong sulphuric acid to a dirty yellow-green solution, in hot acid to a crimson solution with evolution of sulphur dioxide; on diluting with water a brownish solution and black precipitate is obtained. Caustic soda turns the aqueous solution a fine blue colour, hydrochloric acid a red brown.—*Dyes* wool and silk, mordanted with chrome a deep black, fast to acids, light and washing. Used for calico printing with chrome mordant.

Alizarine Blue—

Dioxyanthraquinon quinoline

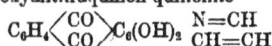

$$C_6H_4\!\!\begin{array}{c}CO\\CO\end{array}\!\!\!>\!\!C_6(OH)_2\quad\begin{array}{c}N=CH\\CH=CH\end{array}$$

obtained by acting with glycerine and sulphuric acid upón *B* nitro alizarine. (1877).—*Dark blue lustrous crystalline powder*, insoluble in water, slightly soluble in hot alcohol with a blue colour; soluble in strong sulphuric acid to a dark bluish red solution; on diluting with water this turns scarlet. Caustic soda turns the alcoholic solution green.—*Dyes* wool and silk mordanted with chrome deep navy blues, very fast to light, acids and washing. Also known as alizarine blue R, alizarine blue GW.

Alizarine Blue S—

(BADISCHE.)—Sodium bisulphite compound of alizarine blue, obtained by acting with the former upon the latter. (1881). *Eng. Pat.* 3603, 1881.—*Sold as a chocolate brown paste or chocolate brown powder*, easily soluble in water with a yellow-brown colour, insoluble in alcohol, soluble in strong sulphuric acid to a fine dark yellow solution; on diluting with water a brown precipitate is obtained. Hydrochloric acid added to the aqueous solution turns it redder, caustic soda turns it a bluish violet.—*Dyes* wool and silk mordanted with chrome deep navy blue, very fast to light, acids and washing.

Alizarine Blue R and GW—

(BAYER.)—Syn. of alizarine blue.

Alizarine Bordeaux B—

(BAYER.)—— Tetroxyanthraquinone. (1890).—*Brown red paste*, insoluble in water, soluble in strong sulphuric acid to a purple solution; on diluting with water it turns brown-red. Soluble in caustic soda to a violet solution.—*Dyes* wool mordanted with chrome dark blue violet; fast to acids, alkalies and soaping. Dyes wool mordanted with alumina dull red; fast to acids, turned blue by alkalies; fast to soaping, which turns the shade a little bluer. Can be used in calico printing.

Alizarine Bordeaux G—

(BAYER.)—— Tetroxyanthraquinone. (1890.)—*Dark brownish yellow paste*, in-

soluble in water, soluble in strong sulphuric acid to a dark crimson solution, on diluting with water a pale reddish brown precipitate is obtained. Soluble in caustic soda to a violet solution.—*Dyes* wool mordanted with chrome dark violet red ; fast to acids, alkalies, and soaping. Dyes wool mordanted with alumina dull wine red ; fast to acids; turned violet with alkalies, turned bluer by soaping Can be used in calico printing.

Alizarine Carmine—

Syn. of alizarine S.

Alizarine Cyanine R—

(BAYER.)—— Pentaoxyanthraquinone. (1890).—*Brown paste*, insoluble in cold water, soluble in strong sulphuric acid to a violet solution, in caustic soda to a blue solution. — *Dyes* wool mordanted with chrome bright blue ; fast to acids, alkalies, washing and light. Dyes wool mordanted with alumina reddish violet, fast to acids ; turned blue by alkalies, fast to washing and light. Applied in calico printing with good results ; gives blue with chrome and reddish violet with alumina mordants. Can be used in calico printing.

Alizarine Indigo Blue—

(BADISCHE.)—Bisulphite of sodium combination of a product formed by the action of sulphuric acid on alizarine green. (1889). — *Eng. Pat.* 15121, 1888. — *Maroon coloured paste*, smelling of sulphur dioxide, soluble in water to a ruby coloured solution, in strong sulphuric acid to a blue solution ; on diluting with water this turns first violet, then crimson, finally red. Caustic soda gives a blue solution.—*Dyes* wool mordanted with chrome bright blues, nearly resembling indigo ; fast to light, acids, alkalies and air.

Alizarine Green—

(BADISCHE.) — (1888). *Eng. Pat.* — *Paste, smelling of sulphurous acid*, appearing black by refected, ruby red by transmitted, light.—*Dyes* wool mordanted with chrome fine deep shades of peacock green ; fast to light, washing and acids.

Alizarine Green—

Syn. of cœrulein and cœrulein S.

Alizarine Orange—

B-Mononitroalizarin

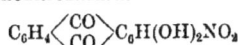

$$C_6H_4\!\!\begin{array}{c}CO\\CO\end{array}\!\!\!>\!\!C_6H(OH)_2NO_2$$

obtained by acting with nitric acid upon alizarine in a nitrobenzol solution. 1876.—*English Patent* 1229, 1876.—

Brownish yellow paste, insoluble in water, soluble in solution of soda to a magenta red solution; on adding acid and zinc powder, this turns first blue then yellow brown; on exposure to the atmosphere the blue colour returns. Soluble in strong sulphuric acid to a yellow brown solution, which on diluting with water gives a pale yellow precipitate. — *Dyes* mordanted fibres; alumina gives orange shades, iron reddish violets, chrome red browns; very fast to light, air, washing, and acids. Known also as alizarine OR, alizarine OG, alizarine N, alizarine orange OR and OG (BAYER) syn. of alizarine orange.

Alizarine S—

Sodium salt of alizarine monosulphuric acid.

$$C_6H_4 \underset{CO}{\overset{CO}{\diagdown\diagup}} C_6H(OH)_2SO_3Na$$

obtained by acting with strong sulphuric acid upon alizarine. (1871).—*Orange yellow powder*, easily soluble in water to a scarlet solution, in alcohol to a yellow solution, in strong sulphuric acid to a scarlet solution; on diluting this with water it turns pale yellow; hydrochloric acid added to the aqueous solution turns it pale yellow; caustic soda turns it violet.— *Dyes* fibres mordanted with alumina fine scarlet reds, with iron dark violets, with chrome Bordeaux reds; very fast to light, acids, washing. Very much used for dyeing wool and silk. Known also as alizarine powder N, alizarine WSL, alizarine carmine.

Alizarine Powder W—

(BAYER.)—Syn. of alizarine S.

Alizarine Violet—

Syn. of gallein.

Alizarine Yellow A—

(BADISCHE.)—Trioxybenzophenone

$$C_6H_5COC_6H_2(OH)_3$$

1889. — *English Patents* 8373, 9427, 9428, 1889.—*Greyish yellow paste*, the colouring matter of which rapidly settles out, leaving a brownish coloured liquor; it is slightly soluble in cold, easily in hot, water to a brownish yellow solution; in alcohol to an amber coloured solution; in strong sulphuric acid to an amber coloured solution, which, on diluting with water, turns paler; soluble in caustic soda to a dark amber solution.—*Dyes* wool mordanted with chrome olive yellows; with alumina, yellow; with iron, olives. Can be used in calico printing with chrome, iron or alumina mordants.

Alizarine Yellow C—

(BADISCHE.)—Gallacetophenone CH₃.CO. $C_6H_2(OH)_3$. (1889.) *Eng. Pat.* 9429, 1889.—*Drab coloured paste*, the colouring matter of which is slightly soluble in water, rapidly settles out on standing, leaving a brownish coloured liquor above it, soluble in boiling water to a faint yellow solution, reprecipitates out on cooling. Soluble in alcohol to a light amber yellow solution. Strong sulphuric acid acts strongly, nitrogen peroxide being given off, and forming a yellow solution, which, on diluting with water, turns pale yellow. Caustic soda forms a dark amber coloured solution.—*Dyes* wool mordanted with chrome brownish yellow; with alumina, olive yellow; with iron, greys. Can be used in calico printing, either with alumina, iron or chrome mordants. Not fast to acids or alkalies. Fast to soaping and light.

Alizarine Yellow GG—

(MEISTER, LUCIUS AND BRUNING.)— *M*-nitrobenzene azo-salicylic acid.

$$C_6H_4NO_2N{:}NC_6H_3OHCOOH$$

obtained by combining *m*-diazonitrobenzene with salicylic acid. (1889).—Pure product crystallises from alcohol in pale yellow needles m.p. 230°C. Very slightly soluble in cold, easily in hot water. The commercial product is in the form of a yellow paste, containing about 20 per cent. of colouring matter.—*Dyes* wool mordanted with chrome greenish yellow, fast to soap; with alumina mordant gives a golden yellow, but this is not fast to washing. Both shades are fast to light and acids. The wool may be mordanted before dyeing, or the mordant may be added direct to the dye-bath. Cotton mordanted with alumina or chrome may be dyed. For printing, the printing colour is made with starch and tragacanth thickening, acetate of chrome, acetic acid and colouring matter, printed and steamed.

Alizarine Yellow R—

(MEISTER, LUCIUS AND BRUNING.)— *P*-nitrobenzene azo-salicylic acid

$$C_6H_4NO_2N : NC_6H_3OH COOH$$

made by combining *p*-nitrodiazobenzene with salicylic acid (1890).—Properties and uses similar to those of alizarine yellow GG, but gives brownish yellows on chrome mordanted fibres.

Alkali Blue, Nicholson Blue—

(SIMPSON, MAULE AND NICHOLSON.) — The sodium salt of triphenylrosaniline or pararosaniline sulphonic acid

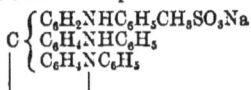

$$C \begin{cases} C_6H_2NHC_6H_5CH_3SO_3Na \\ C_6H_4NHC_6H_5 \\ C_6H_4NC_6H_5 \end{cases}$$

obtained by sulphonating spirit blue (1862.) E. C. Nicholson, English patent, 1st June, 1862.—Gilbee, English patent, 3rd July, 1862.—*Pale or dark blue powder*, slightly soluble in cold water, easily soluble in hot water; on cooling the blue settles out. Soluble in alcohol to a fine blue solution; in strong sulphuric acid to fine brown-red solutions, on diluting with water a blue solution and blue precipitate is obtained. Hydrochloric acid added to the aqueous solution gives a blue precipitate, caustic soda a brown-red solution.—*Dyes* wool and silk by special method fine shades of blue; fast to acids and washing, fairly fast to light.

Alkali Blue D—

(ACTIENGESELLSCHAFT.)—Sodium salt of diphenylamine - monosulphonic acid; obtained by acting with strong sulphuric acid on diphenylamine blue.—*Dark blue powder*, insoluble in cold water, soluble in warm water to a blue solution which smells of diphenylamine. Soluble in strong sulphuric acid to a red-brown solution; on diluting with water a blue precipitate is obtained. Hydrochloric acid flows down a blue precipitate from the aqueous solution. Caustic soda turns the colour grey-violet, on warming this becomes red-brown.—*Dyes* wool and silk in the same way as alkali blue. Good shades of blue are obtained.

Alkali Green—

(BROOKE, SIMPSON AND SPILLER.)—Sodium salt of diphenyl-diamido triphenyl-carbinol sulphonic acid

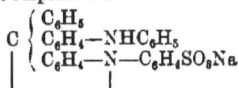

$$C \begin{cases} C_6H_5 \\ C_6H_4-NHC_6H_5 \\ C_6H_4-N-C_6H_4SO_3Na \end{cases}$$

obtained by acting on diphenylamine with benzyl chloride, oxidising and sulphonating the product. Discovered 1877. R. Meldola. *Jour. Chem. Socy.*, 1882. — *Dark green powder*, soluble in water with green colour, on adding hydrochloric acid a green precipitate is thrown down. Caustic soda changes the colour of the solution to brown; strong sulphuric acid dissolves it, forming a magenta red solution; on adding water, a green flocculent precipitate is

formed.—*Dyes* wool and silk in the same way as alkali blue. It is not now used. Known also as viridine.

Amaranth—

(MEISTER, LUCIUS AND BRUNING.) (CASELLA AND CO.)—Sodium or calcium salt of naphthionic acid-azo-*B*-naphthol disulphonic acid

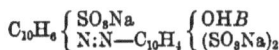

$$C_{10}H_6 \begin{cases} SO_3Na \\ N:N-C_{10}H_4 \end{cases} \begin{cases} OHB \\ (SO_3Na)_2 \end{cases}$$

obtained by combining naphthionic acid and *B*-naphthol disulphonic acid R. Discovered in 1878.—*Red brown powder*, soluble in water with a magenta-red colour, which is turned darker on addition of caustic soda. Soluble in concentrated sulphuric acid with a violet colour, changing on addition of water to the original red.—*Dyes* wool dark crimson in acid baths. One of the fastest colours known. Known also as fast red D (BADISCHE), claret red S, œnanthinine; Bordeaux S. (BER. ANIL), azoacid rubine 2B (DAHL.).

Amethyst—

(KALLE AND CO.)—Diethylparaamido phenyl-diethylamido phenazonium chloride.

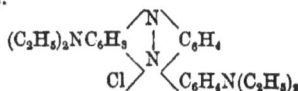

$$(C_2H_5)_2NC_6H_3 \underset{Cl}{\overset{\diagup N \diagdown}{\underset{\diagdown N \diagup}{}}} \begin{matrix} C_6H_4 \\ C_6H_4N(C_2H_5)_2 \end{matrix}$$

obtained by oxidising equal molecules of diethylparaphenylene diamine, diethylaniline and aniline.—*Black-grey powder*, soluble in water to a red-violet solution; in alcohol to a crimson solution, having a brown-red fluorescence; in strong sulphuric acid to a fine green solution; on diluting with water this turns first blue then violet. Hydrochloric acid added to the aqueous solution turns it blue, caustic soda has little action.—*Dyes* cotton mordanted with tannin and tartar emetic, wool and silk in neutral baths violets; silk has a red fluorescence. Not now in use.

Amines—

A group of bodies derived from ammonia NH_3 by the substitution of one or more of the hydrogen atoms by equivalent quantities of organic radicles; thus there is methylamine CH_3NH_2, phenylamine $C_6H_5NH_2$, and naphthylamine $C_{10}H_7NH_2$. Generally they can be obtained by acting with ammonia on the chlorides, iodides, or bromides of the organic radicles, such as phenyl, ethyl, naphthyl, benzyl, etc. Primary amines are those which have only one hydrogen of the ammonia replaced (see those quoted above). Secondary amines have two hydrogens replaced, as for example : diphenylamine $N(C_6H_5)_2H$, methylethylamine $NC H_5C_2H_5H$. Tertiary amines have all the hydrogen atoms replaced, as for example : trimethylamine. The following formulæ show the relationship of each of these groups of amines to ammonia :—

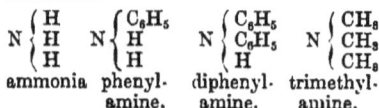

$$N \begin{cases} H \\ H \\ H \end{cases} \quad N \begin{cases} C_6H_5 \\ H \\ H \end{cases} \quad N \begin{cases} C_6H_5 \\ C_6H_5 \\ H \end{cases} \quad N \begin{cases} CH_3 \\ CH_3 \\ CH_3 \end{cases}$$

ammonia   phenyl-   diphenyl-   trimethyl-
        amine.   amine.   amine.

Diamines are derived from diatomic organic radicles, and contain the group twice NH_2 as phenylene diamine $C_6H_4(NH_2)_2$, toluylene diamine $C_7H_6(NH_2)_2$.

All these bodies are basic bodies, and combine directly with acids to form bodies perfectly analogous to the salts of ammonia thus :—

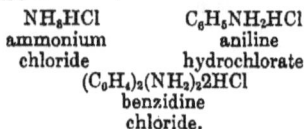

NH_3HCl $C_6H_5NH_2HCl$
ammonium   aniline
chloride   hydrochlorate
      $(C_6H_4)_2(NH_2)_22HCl$
         benzidine
         chloride.

Like the ammonium compounds, these amine salts are volatile, and are decomposed by caustic soda or potash, the free amine being liberated. The group NH_2 characteristic of the primary amines is known as the amido group, and all bodies containing it are called amido compounds, and they are tolerably numerous. The amines are not colouring matters, nor do they possess any marked colour, nor are they in a sense colour producers. The introduction of the NH_2 group into any compound does not convert that compound into a colouring matter ; but the amines are valuable as being the bases from which colouring matters can be obtained by various means that do not require specifying here.

It may be noted here that the amines of the aromatic hydrocarbons, benzene, naphthalene, etc., when treated with nitrous acid in the cold, become azotised and converted into bodies of the types :—

$C_6H_5N:NC_6H_5$ $C_6H_5N:NCl$
azobenzene.   diazobenzene chloride.

(See AZO-COLOURING MATTERS). The amines of the fatty hydrocarbons, ethane, propane, ethene, etc., do not undergo this reaction, but are converted into the corresponding alcohols. The amines of the aromatic hydrocarbons are generally produced by reducing the nitro-compounds :—Nitrobenzene, nitro-toluene, nitro-naphthalene, with iron or zinc in the presence of an acid. The hydrogen liberated by the reaction of the metal and acid react on the nitro compound, and convert it in the amine thus :—

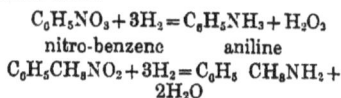

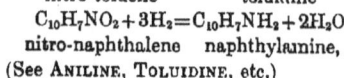

$$C_6H_5NO_3 + 3H_2 = C_6H_5NH_3 + H_2O_2$$
nitro-benzene   aniline
$$C_6H_5CH_3NO_2 + 3H_2 = C_6H_5 CH_3NH_2 + 2H_2O$$
nitro-toluene   toluidine
$$C_{10}H_7NO_2 + 3H_2 = C_{10}H_7NH_2 + 2H_2O$$
nitro-naphthalene   naphthylamine,

(See ANILINE, TOLUIDINE, etc.)

Aniline Mauve—

Syn. of mauveine.

Aniline—

Phenylamine $C_6H_5NH_2$—This important body was first obtained from indigo in 1826 by Unverdorben ; it is present in small quantities in coal tar. On the large scale nitrobenzene is treated with iron filings and hydrochloric acid in large iron stills ; the operation takes about 8 hours ; then quicklime is added, which liberates the aniline from the acid, and by means of steam it is distilled over. When pure, aniline is a colourless oily liquid, having a peculiar and rather disagreeable odour. It boils at 182° C. It is a trifle heavier than water, its specific gravity being 1·0265. It is slightly soluble in water (1 in 31), readily soluble in alcohol, ether, and other solvents. It unites with acids, forming the aniline salts, such as aniline hydrochlorate $C_6H_5NH_2HCl$; this is the only one of the salts having a commercial value ; they are all soluble in water. Aniline when exposed to the air and light soon turns to a brownish colour, this change being due to some impurity that is not easily got rid of. Commercially, several qualities of "aniline oil" are distinguished :—Aniline oil for blue and black ; aniline oil for red ; aniline oil for saffranine ; toluidine. The first is almost pure aniline, contains 90 per cent. at least, which distils up to 182' C. Aniline oil for red is a mixture of about equal parts of aniline and toluidine, and is largely used for making magenta. Aniline oil for saffranine contains a good deal of toluidine. Heating aniline with glacial acetic acid converts it into acetanilide.

$$C_6H_5NHC_2H_8O.$$

Aniline is used for making magenta, saffranine, blues, induline, quinoline, etc.

Aniline Black—

This differs from all other aniline colours with one exception—naphthylamine violet – in that it must be produced direct on the fibre, although the black may be obtained in the free state in the form of a black powder, which, however, is so very insoluble that it cannot be used as a dye, and only finds a limited application as a pigment in calico printing and painting. The composition and constitution of aniline black have not yet been properly determined; it is supposed to be the hydrochloride of a base having the formula—

$$C_{30}H_{25}N_5$$

although Liechti and Suida assign to aniline black the formula—

$$C_{18}H_{16}ClN_3;$$

this uncertainty probably arises from the fact that the composition of an aniline black will greatly depend upon the method and materials by and from which it has been made. Aniline black is produced by the weak oxidation of aniline and toluidine, such oxidation being brought about by the use of oxidising agents of various kinds—bichromate of potash, sulphate of copper, ferric chloride, copper sulphide, and other salts. Too energetic oxidation must be avoided, as then quinone is formed. Aniline black was first patented by John Lightfoot, 1863, No. 151, since which time many patents have been and are being taken out. Lightfoot used the black both for dyeing and printing; for dyeing it was made with

1gal. water.
4oz. chlorate of potash.
8oz. aniline, previously mixed with 8oz. hydrochloric acid, 32° Tw.
1 part acetic acid.
8oz. measures chloride of copper, 88° Tw.
4oz. sal ammoniac.

The fabric was steeped in this, wrung out, and aged for two or three days at 70° F.; then passed through ammonia. For printing the following colour was made :—

1gal. starch paste (1lb. starch).
4oz. chlorate of potash.
8oz. aniline, previously mixed with 8oz. hydrochloric acid, 32° Tw.
4oz. measure perchloride of copper, 88° Tw.
8oz. sal ammoniac.

The fabrics are printed, aged for three nights, and raised in alkaline bath. He also uses red prussiate of potash in place of the chloride of copper, in which case he uses 4oz. to 6oz. of the salt to 1lb. aniline salt, adding 2oz. oxalic acid to a gallon of printing colour. Lauth substituted sulphide of copper for the chloride, which exerts a destructive action on the doctors of the printing

machine. Ammonium vanadate is a powerful black producer; one part will lead to the production of 50,000 parts of black, and hence it is much used. A simple method of dyeing consists in working the cloths in a dye-bath composed of 2lb. aniline, 2lb. hydrochloric acid, dissolved in 4 gal. of water, to which is then added 8lb. bichromate of potash dissolved in water. The goods are worked in this until the black is fully developed, then it is washed and dried; there is, however, some loss of colouring matter by this process. There is scarcely any doubt but what the best aniline blacks are produced by such methods as require ageing. Such a black is produced from a mixture of chlorate of soda, aniline hydrochlorate, and potassium ferrocyanide, working the cloth or fibre in this, ageing at about 100° F. until the cloth has acquired a deep olive-green, then developing the black in a bath of bichromate of potash. The great fault of many processes, especially when the chlorates of potash and soda are used, is that during the subsequent ageing there is a development of chlorine, and this, acting on the fibre, tenders it very much. In calico printing aniline black is largely used, and has almost replaced the logwood blacks which were used previous to its advent. Lightfoot's colour has already been given. Lauth's colour is made with 10 litres starch paste, 350grm. chlorate of potash, 300grm. copper sulphate, 300grm. sal ammoniac, and 800grm. of aniline hydrochlorate. The cloths are printed, aged for two days at 80° to 90° F., then passed through an alkaline bath. Pieces printed with aniline black containing chlorates cannot be steamed, as there is then great tendency to tendering, consequently such aniline blacks cannot be combined with steam colours. In such cases what is called a steam black is made, in which some of the chlorate is replaced by potassium ferrocyanide; after printing the cloths are steamed. Discharges on aniline blacks cannot be obtained, so that where whites on blacks are wanted it is necessary to use reserves or resists; these are always more or less alkaline in character, and act by preventing the development of the black. For further details reference must be made to works on calico printing. Aniline black is characterised by its fastness to acids, alkalies, soaping and light, when it has been properly developed. Some blacks have a tendency to turn green by exposure to air and light; this is generally due to the black not being sufficiently oxidised. It is of interest to note that it has been observed that blacks obtained from pure aniline are more liable to green than blacks got from toluidine or xylidine, or a mixture of those bases. Aniline, however, gives the finest black, para-toluidine gives a brownish black, ortho-toluidine gives a blue-black, and does not require chroming to develop the full shade, and xylidine a brown-black; consequently most dyers use pure aniline for the production of their blacks. Space does not admit of a fuller treatment of the subject in the present work.

Aniline Blue Spirit, Soluble—

Syn. of opal blue.

Aniline Brown—

Syn. of Bismarck brown.

Aniline Green—

Syn. of aldehyde green.

Aniline Orange—

Syn. of phosphine and of Victoria yellow.

Aniline Purple—

An old name for mauveine.

Aniline Red—

Syn. of magenta.

Aniline Rose—

Syn. of saffranine.

Aniline Violet—

An old name for mauveine.

Aniline Yellow—

Hydrochloride of amidoazo benzene

$$C_6H_4NH_2HClN : NC_6H_5$$

obtained by treating diazoamido benzene with hydrochloric acid. Discovered by Mene in 1861. Introduced by Simpson, Maule, and Nicholson in 1863. Dale and Caro. *Eng. Pat.* 3307, 1863. Witt and Thomas, *Chem. Socy. Journ.* 1883, p. 112 ; Friswell and Green, *Chem. Socy. Journ.* 1885, p. 917 ; 1886, p. 746.—*In steel blue lustrous crystals,* soluble in water with a yellow colour, in strong sulphuric acid to a brown solution. On diluting with water, this turns red. Acids turn the colour of the aqueous solution red. On warming the solution, the free base precipitates.—*Dyes* wool and silk red in acid baths. On washing with water, the dyed fibre turns yellow. It is not fast, the colour easily volatilising when heated ; it is therefore not now used as a colouring matter, although used as a base for obtain-

ing other colours. Known also as spirit yellow.

Anisidin Scarlet—

Syn. of anisol red.

Anisol—

$$C_6H_5OCH_3 \ methyl \ phenate,$$

is used to make some colouring matters ; it is made by heating potassium phenate with iodide of methyl. It is a liquid having a pleasant aromatic odour. Nitric acid converts it into nitranisol,

$$C_6H_4O(NO_3)CH_3 ;$$

on reduction this yields anisidin

$$C_6H_4ONH_2CH_3.$$

Dianisidin formed from the dinitranisol has the

formula $\begin{array}{c} C_6H_3OCH_3NH_2 \\ | \\ C_6H_3OCH_3NH_2 \end{array}$ and is used in

making benzoazurine, etc.

Anisol Red—

(BADISCHE.)—Sodium salt of o-anisidin-azo-*B*-naphthol monosulphonic acid

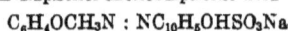

$$C_6H_4OCH_3N : NC_{10}H_6OHSO_3Na$$

obtained by diazotising o-anisidin and treating with *B*-naphthol monosulphonic acid S. (1878.) *Eng. Pat.* 4726, 1878.— Griess—*Brown-red powder,* easily soluble in water, with a cherry-red colour, and in strong sulphuric acid with a magenta colour ; addition of water turns the colour to a cherry red ; adding acids to the aqueous solution produces no change. Caustic soda turns the colour yellowish red. —*Dyes* wool and silk a fiery red in acid baths. Known also as anisidin scarlet.

Anthracene—

$$C_{14}H_{10}.$$

An aromatic hydrocarbon found in coal-tar. Can be obtained in colourless tabular crystals having a fine blue fluorescence, generally got by subliming the crude anthracene of the tar distiller by the aid of superheated steam in the form of flakes. It is sparingly soluble in alcohol and ether, freely soluble in benzene, insoluble in water, melts at 213° C., distils at a temperature of about 360° C. An isomeric modification known as paranthracene exists, which is characterised by being insoluble in benzene and melting at 244° C. By the action of fuming sulphuric acid anthracene is converted into sulphonic acids ; by oxidation with nitric or chromic acids anthracene is converted into anthraquinone, $C_{14}H_8O_2$. The constitution of anthracene is shown in the formula—

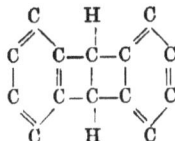

it contains two complete benzene nuclei, united by the group $\begin{array}{c}=\text{CH}\\ |\\ =\text{CH}\end{array}$ It is of interest as forming the basis or starting point for the manufacture of alizarine.

Anthracene Brown—

Syn. of anthragallol.

Anthracene Green —

Syn. of cœrulein and cœrulein S.

Anthracene Violet—

Syn. of gallein.

Anthragallol. Anthracene Brown—

Trihydroxyanthraquinone

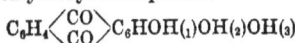

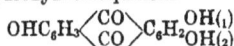

$$C_6H_4\!\!\begin{array}{c}CO\\ \diagdown\\ CO\end{array}\!\!C_6HOH_{(1)}OH_{(2)}OH_{(3)}$$

formed by heating a mixture of gallic acid, benzoic acid, and sulphuric acid together. Is isomeric with anthrapurpurin, crystallises in yellow needles, subliming at 290° C., slightly soluble in water, readily in alcohol, and in alkalies to a green solution. The commercial product is a dark brown paste.—It dyes wool mordanted with bichrome a fine brown, which is very fast to light, milling, and acids.

Anthrapurpurin—

Trihydroxyanthraquinone

$$OHC_6H_3\!\!\begin{array}{c}CO\\ \diagdown\\ CO\end{array}\!\!C_6H_2\begin{array}{c}OH_{(1)}\\ OH_{(2)}\end{array}$$

This is a most important colouring matter. It is found in the crude alizarine melt, and can also be prepared from other anthracene products. Anthrapurpurin (also called isopurpurin) crystallises in orange-coloured needles, melting above 330° C.; soluble in alkalies, with a violet colour. It dyes mordanted fibres in a similar manner to alizarine; the reds with alumina are, however, purer and less blue, the purples are bluer, and the blacks stronger. It is the most used variety of commercial alizarine for Turkey red dyeing. The commercial article is in the form of a yellow-brown paste, containing 20 per cent. of actual colouring matter, known as alizarine GD (BADISCHE), alizarine RX (MEISTER), alizarine SX (BAYER), alizarine RF (LEVERKUS).

Archil Red A—

(BADISCHE.)—Sodium salt of amidoazoxyleneazo-B-naphthol disulphonic acid

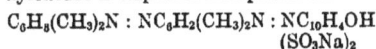

$$C_6H_3(CH_3)_2N : NC_6H_2(CH_3)_2N : NC_{10}H_4OH\\ (SO_3Na)_2$$

from amidoazoxylene, by diazotising and combining with B-naphthol disulphonic acid R. (1880.)—*Dark brown powder*, soluble in water with a wine red colour; in strong sulphuric acid to a dark blue solution; on diluting with water a red-brown flocculent precipitate is obtained; addition of hydrochloric acid to the aqueous solution produces a brown-red precipitate, while caustic soda turns the colour brown. —*Dyes* wool in acid baths a bluish red, fast to acids, light and washing.

Archil Substitute—

Syn. of naphthionic red.

Atlas Scarlet No. 3—

(BROOKE, SIMPSON AND SPILLER.) — Sodium salt of sulphoxyleneazo-B-naphthol disulphonic acid

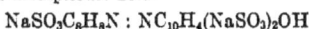

$$NaSO_3C_6H_3N : NC_{10}H_4(NaSO_3)_2OH$$

obtained by acting with diazotised xylidine sulphonic acid on B-naphthol disulphonic acid. Meldola. *English Patent*, 1864; May 10, 1879.—*Chem. News.*, vol. xlii., p. 60.—*A scarlet powder*, soluble in water to scarlet solution; in alcohol to amber-coloured solution; in acetic acid to scarlet solution; in strong sulphuric acid to dark violet solution, which on diluting with water turns to a salmon colour. Hydrochloric acid added to the aqueous solution throws down a scarlet precipitate. Caustic soda turns the colour of the solution yellower.—*Dyes* wool and silk from acid baths bright full shades of scarlet, fast to dilute acids; turned purplish by strong acids and caustic soda. Not fast to boiling in soap or to light.

Auramine—

(BADISCHE.)—Socy. Chem. Ind., Basle. Hydrochlorate of amido-tetramethyl diamido-diphenyl methane.

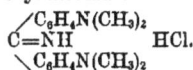

$$C\!\!\begin{array}{c}\diagup C_6H_4N(CH_3)_2\\ =NH\\ \diagdown C_6H_4N(CH_3)_2\end{array}\quad HCl.$$

(1883.) Caro and Kern. *English Patents*, 5512, 1884; 5741, 1884.—*Sulphur yellow powder*, slightly soluble in cold water, easily soluble in hot water to a pale yellow solution, in strong sulphuric acid to a colourless solution, on diluting with water the colour returns; soluble in alcohol; addition of hydrochloric acid to the aqueous solution causes a partial loss of colour. On heating, the colouring matter decomposes with formation of ammonium chloride and tetramethyl diamidobenzophenone. On adding caustic soda to the aqueous solution the base is thrown down as a white precipitate, soluble in ether and alcohol. These solutions take a yellow colour on the addition of acetic acid.—*Dyes* cotton mordanted with tannin a sulphur yellow, fast to light, and light washing. Dyes wool in neutral bath, and silk in a neutral bath fine shades of greenish yellow. Auramine is sold in three brands: Auramine O, which is the nearly pure colouring, and Auramine I. and II., which contain various proportions of dextrine.

Aurantia —

(BER. ANIL.)—The ammonia salt of hexanitrodiphenylamine

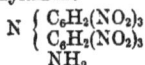

$$N \begin{cases} C_6H_2(NO_2)_3 \\ C_6H_2(NO_2)_3 \\ NH_2 \end{cases}$$

obtained by treating diphenylamine with nitric acid. Discovered in 1873. Kopp.— *Red brown crystals*, soluble in water with orange-yellow colour; acids precipitate the free acid as a sulphur yellow flocculent precipitate, the solution becoming almost colourless; alkalies darken and redden the colour; metallic salts turn the colour brown.—*Dyes* wool and silk a fine orange-yellow in slightly acid baths. Known as Imperial yellow.

Aureosin—

A colouring matter obtained by acting with hypochlorous acid upon fluorescin, not now in use.—*In yellow brown powder or pieces*, very slightly soluble in water, easily soluble in caustic soda to a brown-yellow solution which has a green fluorescence, soluble in strong sulphuric acid to a yellow solution; on diluting with water an orange-yellow precipitate is obtained.

Aurine —

A mixture of aurine, methylaurine and pseudorosolic acid, obtained by heating phenol with oxalic acid and sulphuric acid.

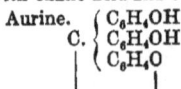

$$\text{Aurine. C.} \begin{cases} C_6H_4OH \\ C_6H_4OH \\ C_6H_4O \end{cases}$$

1834, Runge from crude coal-tar acids; 1859, Persoz from phenol and oxalic acid; 1866, Caro and Wanklyn, from rosaniline. *Chem. News*, 1866, vol. xiv., p. 3. Caro, *Phil. Mag.*, 1866, p. 125.—*Yellow brown pasty masses, with a dark green metallic lustre*, insoluble in water, soluble in alcohol with a yellow colour, in strong

sulphuric acid to yellow solution; hydrochloric added to the alcoholic solution has no action; caustic soda turns to a bright crimson. Used for colouring spirit lacquers and varnishes, and for making other colours.

Aurine R—

Syn. of pæonine.

Aurotine—

(CLAYTON ANILINE COMPANY.)—Sodium salt of tetranitrophenolphthalein

$$C_{20}H_8(ONa)_2O_2(NO_2)_4$$

obtained by nitrating phenolphthalein. (1889.) *English Patent* 3441, 1889.— *Orange yellow powder*, soluble in water and alcohol to a yellow solution, and in strong sulphuric acid to a yellow solution, on diluting with water it becomes colourless; acids added to the aqueous solution discharge the colour, caustic soda has no action.—*Dyes* wool from an acid bath or on a chrome mordant orange-yellow, fairly fast to light and washing; acids discharge the colour, alkalies have no action.

Azaleine—

Syn. of magenta. *English Patent* 2800, 1859.

Azarin S—

(MEISTER, LUCIUS AND BRUNING).—Ammonia salt of dichloramidophenol-hydrazo-*B*-naphtho sulphonic acid

$$C_6H_2Cl_2OHNH—NSO_3NH_4C_{10}H_6OH$$

made by diazotising dichlorphenol, combining with *B*-naphthol and treating with ammonia bisulphite. (1883.) *English Patent* 5767, 1883.—The process of dissolving insoluble colouring matters by means of bisulphites was discovered by Prudhomme about 1879, and since then has been very widely applied; notably in the case of the alizarin series of colours. —*Yellow paste, smelling of sulphur dioxide*, only slightly soluble in water to a yellow solution, in strong sulphuric acid to a crimson-red solution, on diluting with water a red-brown precipitate is obtained. Adding hydrochloric acid to the aqueous solution produces an orange-yellow precipitate; caustic soda gives a violet precipitate, soluble with a red colour on boiling. Used principally for calico printing.

Azines—

This name is given to a group of bodies which form the base of many useful colouring matters, such as the safranines;

the characteristic feature of azines is their containing the group of elements—

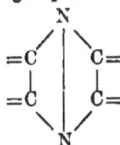

$$=C\diagup\!\!\!\!\overset{N}{\underset{N}{\diagdown}}\!\!\!\!\diagup C=$$
$$=C\qquad\qquad C=$$

The carbons form part of bivalent radicles belonging either to the fatty or aromatic series. This characteristic group possesses what Witt calls "chromophoric properties," that is, capable of imparting colouring properties when present in a compound; every azine is a chromogene, *i.e.*, a colour producer; their amido and oxy derivatives are all colouring matters. The amido derivatives were first prepared by Witt, and named by him eurhodines (*Chem. Socy. Journal*, 1886, p. 391), while the hydroxy derivatives are called eurhodols. The saffranines also belong to the azines, and contain the group—

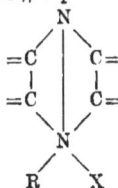

$$=C\diagup\!\!\!\!\overset{N}{\underset{N}{\diagdown}}\!\!\!\!\diagup C=$$
$$=C\qquad\qquad C=$$
$$R\qquad\qquad X$$

where R = any monovalent organic radicle; ethyl C_2H_5, phenyl C_6H_5, benzyl C_7H_7, etc., and X, Cl, NO_2, or similar acid group. The colouring matters derived from azines are weak basic bodies, dyeing on cotton mordanted with tannin and tartar emetic, although some of them have also a weak affinity for the cotton fibre. The azines form salts with mineral but not with organic acids; these are usually of a red or orange colour, and decomposed by water, the free azine being liberated. They are soluble in strong sulphuric acid with intense colourations, changing instantly on diluting with water, which, when in excess, precipitates the free azine. The azines themselves are not colouring matters, and become such only when combined with amido or hydroxy groups. The colouring matters which are of practical importance are the eurhodines, eurhodols, saffranines (which see). For further information Thorpe's " Dictionary of Applied Chemistry," vol. i., p. 234, article AZINES, may be consulted. Cotton requires a mordant of tannin and tartar emetic or tin crystals, and excellent results are obtained. Wool is dyed in a neutral bath with Glauber's salt only just under the boil, the azines having strong affinity

for this fibre. Silk is dyed in a plain bath just hand warm, or better in a bath of old boiled off liquor. In calico printing the azines are applied with a thickening containing tannin and acetic acid, fixing after printing, and steaming in a bath of tartar emetic.

Azine Green G—

(LEONHARDT.)—Probably,

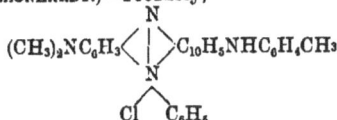

$$(CH_3)_2NC_6H_3\diagup\!\!\!\!\overset{N}{\underset{N}{\diagdown}}\!\!\!\!\diagdown C_{10}H_5NHC_6H_4CH_3$$
$$Cl\quad C_6H_5$$

(1890.) *English Patent* 3090, 1890.— *Blackish grey powder*, soluble in water to a dark blue-green solution, in strong sulphuric acid to a red-brown solution, on diluting with water a green solution is obtained. Hydrochloric acid turns the aqueous solution a little bluer, caustic soda decolourises it.—*Dyes* cotton mordanted with tannin and antimony deep blue-green, fast to acids, light and soaping.

Azine Green B—

(LEONHARDT.)—(1890.) *English Patent* 3090, 1890.—*Black powder*, slightly soluble in cold, easily in hot water, to a dark green solution, in acetic acid to a dark blue solution, in strong sulphuric acid to a dark brown solution. Hydrochloric acid turns the aqueous solution dark blue; caustic soda gives a dark brown precipitate.—*Dyes* cotton mordanted with tannin and antimony deep blue-green, fast to acids, light and washing.

Azo Black—

(MEISTER, LUCIUS AND BRUNING.)—Syn. of blue black (BADISCHE).

Azo Blue—

(BAYER.) (ACTIENGESELLSCHAFT.)

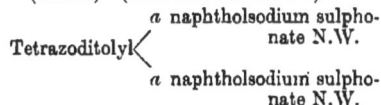

$$\text{Tetrazoditolyl}\diagdown\!\!\!\!\diagup\begin{array}{l}a\ \text{naphtholsodium sulphonate N.W.}\\ a\ \text{naphtholsodium sulphonate N.W.}\end{array}$$

(1885.) *English Patent* 9510, 1885.— *Blueish black powder*, readily soluble in water to a violet solution, in strong sulphuric acid to blue solution; on diluting with water a violet precipitate falls. Hydrochloric acid gives a dark blue precipitate, caustic soda turns the aqueous solution crimson.—*Dyes* cotton in a neutral soap bath a violet shade of blue, fairly fast to light, fast to acids and soap; alkalies redden it. Dyes wool a red shade of blue. Dyes silk a grey from a neutral soap bath. Not very suitable for wool or silk.

Azo Carmine—

(BADISCHE.)—Sodium salt of the disulphonic acid of phenylrosinduline. (1889.) *English Patent* 15,259, 1888.—*A paste of a red colour with a peculiar brilliant golden fluorescence*; on standing the colouring matter separates out. Insoluble in cold, but freely in hot water to a red solution, from which the colouring matter again separates on cooling. Strong sulphuric acid dissolves it to a green coloured solution; on adding water the red colour is restored.—*Dyes* wool and silk in acid baths a fine crimson red, fairly fast to acids, light and soaping.

Azococcin 7B—

(BER. ANIL.)—Sodium salt of amidoazobenzene azo-*a*-naphthol monosulphonic acid N.W.

$$C_6H_3N : NC_6H_4N : NC_{10}H_5 \begin{cases} OH \\ SO_3Na \end{cases}$$

obtained by combining amidoazobenzene with *a*-naphthol monosulphonic acid N.W. Discovered in 1883. Witt.—*Brown powder*, slightly soluble in water with magenta colour; on adding hydrochloric acid a brown-red precipitate is formed, on adding caustic soda a red-violet precipitate is thrown down, soluble in water. Dissolves in strong sulphuric acid with a blue-violet colour; on diluting with water a brown-red precipitate is formed. – *Dyes* wool in acid bath fairly fast shades of red.

Azococcin G—

Name applied formerly to tropæolin 0000.

Azococcin 2R –

(BER. ANIL.)—Sodium salt of xylene azo-*a*-naphthol sulphonic acid.

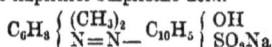

$$C_6H_3 \begin{cases} (CH_3)_2 \\ N=N- \end{cases} C_{10}H_5 \begin{cases} OH \\ SO_3Na \end{cases}$$

obtained by combining xylidine with *a*-naphthol monosulphonic acid N.W. Discovered in 1882. Witt.—*Red brown powder*, slightly soluble in water; on adding hydrochloric acid a brown-red flocculent precipitate is obtained; caustic soda changes the colour to brown-yellow. Soluble in concentrated sulphuric acid with a magenta colour; on diluting with water a brown-red precipitate is obtained.— *Dyes* wool in acid bath red, moderately fast.

Azo Colouring Matters—

These are a very large and important group of colouring matters, comprising many of the most useful dyestuffs at the command of the dyer. The characteristic feature of this group is that they contain one or more of the azo group—N : N—of nitrogen atoms which link together aromatic and other radicles. The earliest known member of this group is azobenzene $C_6H_5N : NC_6H_5$. Discovered by Mitscherlich in 1834; the earliest known colouring matter was ANILINE YELLOW (which see), which is the hydrochloride of amidoazo-benzene, introduced in 1863 by Simpson, Maule and Nicholson. The foundation of all modern discoveries of azo colouring matters was laid by Griess in 1858, when he prepared diazoamidobenzene from nitrous acid and aniline, and again in 1866, when he prepared the first acid azo compound oxy-azobenzene $C_6H_5N : NC_6H_4OH$, and by Kekule and Hidegh when they found out the re-action in alkaline solutions which takes place between diazo compounds and phenols, amines, etc. The first acid azo colours were introduced in 1879, patented by Griess in 1878, and since then they have followed in rapid succession. It is not intended here to do more than give the merest outline of the general method of preparation and composition of these azo bodies. For fuller details reference can be made to Watts' "Dictionary of Chemistry," new edition; to Thorpe's "Dictionary of Applied Chemistry," or to Benedikt and Knecht's "Chemistry of Coal Tar Colours." When aniline hydrochlorate is treated in the cold with a solution of sodium nitrite and hydrochloric acid, diazobenzene chloride is formed thus—

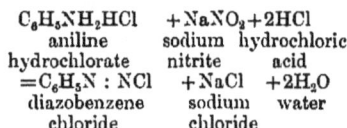

$$C_6H_5NH_2HCl \quad +NaNO_2+2HCl$$
aniline   sodium   hydrochloric
hydrochlorate   nitrite   acid
$$=C_6H_5N : NCl \quad +NaCl \quad +2H_2O$$
diazobenzene   sodium   water
chloride   chloride

When, as in the above substance, the azo group—N : N—is combined on the one hand with an organic radicle, and on the other with an acid radicle like Cl or NO_3, or NO_2, the compounds so produced are called "diazo" bodies. When the group is combined with two organic radicles "azo" compounds are obtained. The diazo compounds are very unstable compounds, readily decompose, especially under the application of heat, hence the necessity of working as cold as possible. When they are treated with alkaline solutions of phenols, there is interaction, and colouring matters are formed; when the phenols themselves are used these are insoluble in water, and are useless as dyestuffs, but methods of producing these direct on the fibre have been patented by Messrs. Read Holliday and Sons, and give satisfactory results. If these insoluble colouring matters are sulphonated, or if the sulphonic acids of the phenols are used, then the colouring matters can be obtained in a soluble form, and in this state they are sent out for use as dyestuffs. The parent compounds from which the azo-colouring matters are derived are azobenzene $C_6H_5N : NC_6H_5$, azotoluene $C_7H_7N : NC_6H_5$, benzeneazonaphthalene $C_6H_5N : NC_6H_5$, azonaphthalene $C_{10}H_7N : NC_{10}H_7$, and their homologues. These are devoid of colouring powers; their amido derivatives have basic properties increasing with the number of amido groups, and as these increase there is a regular gradation of colour; thus in the azobenzene derivatives, amido-azobenzene (Aniline Yellow)

$$C_6H_5N : NC_6H_4NH_2$$

is yellow and weakly basic; diamido-azobenzene (Chrysoidine)

$$C_6H_5N : NC_6H_3(NH_2)_2$$

is orange and more basic; while triamido-azobenzene (Bismarck Brown)

$$NH_2C_6H_4N : NC_6H_3(NH_2)_2$$

is reddish brown and strongly basic. These have considerable affinity for animal fibres, and are capable of dyeing vegetable fibres mordanted with tannin. The hydroxy derivatives, such as oxyazobenzene

$$C_6H_5N : NC_6H_4OH$$

insoluble in water but soluble in alkalies, are strong colouring matters, which are, however, not readily applicable in dyeing; hence they are not much used. The sulphonated compounds are produced either by sulphonating the colouring matter, or preferably by using the amido or phenolsulphonic acids to diazotise or conjugate. The alkaline salts of the sulphonic acid so produced are soluble in water, have a strong affinity for animal fibres, which they dye without mordant, but have no affinity for vegetable fibres, especially the simple azo compounds. As in the amido-azo compounds noted above, there is a regular gradation of colour shown by the various members of an homologous series; thus aniline and its homologues with naphthol disulphonic acid give the following series of colouring matters:— Benzene - azo - beta - naphthol disulphonic acid

$$C_6H_5N : NC_{10}H_4OH(HSO_3)_2$$

Orange. (Orange GT). Toluene-azo-beta-naphthol disulphonic acid

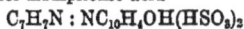

$$C_7H_7N : NC_{10}H_4OH(HSO_3)_2$$

Reddish Orange. (Ponceau GT). Xylene-azo-beta-naphthol disulphonic acid

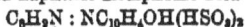

$$C_8H_9N : NC_{10}H_4OH(HSO_3)_2$$

Scarlet. (Scarlet G). Cumene-azo-beta-naphthol disulphonic acid

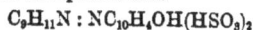

$$C_9H_{11}N : NC_{10}H_4OH(HSO_3)_2$$

Bluish Scarlet. (Ponceau 8R). The tint is also influenced by isomerism of one or the other of the components; thus there are several *B*-naphthol disulphonic acids. Two of these show well marked differences, and are distinguished as the G acid and R acid, the former giving colouring matters of decidedly yellower tints than the R acid, although they are identical in composition. Many examples of this variation of tint produced by the use of isomeric bodies will be found scattered through the numerous group of azo colouring matters. Many azo colouring matters contain the azo group N:N more than once; these are known as secondary or disazo bodies. The earliest colouring matters belonging to this group which were made were Biebrich Scarlet,

$$HSO_3C_6H_4N:NC_6H_3(HSO_3)N:NC_{10}H_6OH$$

sulpho - benzene - azo - sulpho-benzene-azo-*B*-naphthol ; and the Croceine Scarlets,

$$C_6H_4(SO_3H)N:NC_6H_4N:N C_{10}H_5(SO_3H)OH$$

p - sulpho - benzene - azo - benzene - azo - *B* - naphthol sulphonic acid. Many of these secondary azo dyes have an affinity for vegetable fibres, and can therefore be used to dye cotton, etc. Another class of secondary azo bodies are derived from bivalent organic radicles like diphenyl

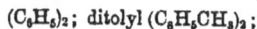

$$(C_6H_5)_2 ; \text{ ditolyl } (C_6H_5CH_3)_2 ;$$

these are generally know as tetrazo compounds, and as a rule the dyestuffs have a strong affinity for the cotton fibres (see BENZIDINE COLOURING MATTERS). A few tertiary or triazo colouring matters are known, such as *B*-oxy-naphthalene-*p*-azo-benzene-azo-*m*-xylene-azo-*B*-naphthol,

$$HOC_{10}H_6N:NC_6H_4N:NC_6H_2(CH_3)_2N:NC_{10}H_6OH$$

which dyes wool or silk dull violet. It may be mentioned that numerous azo colouring matters are known that will dye wool or silk, but which owing to the fugitive character, or dulness, or some other disadvantages of the tints produced are not offered as technical products. The azo colours dissolve with characteristic colours in strong sulphuric acid, affording to some extent a rough test for these bodies. The primary azo colours containing only one azo group N:N dissolve with a yellow or orange or red colour. Secondary azo colours dissolve with various colours dependent upon the way in which the HSO_3 group is combined with the other radicles of the colour. When these are contained in the benzene groups only, as in Biebrich Scarlet,

$$HSO_3C_9H_4N : NC_6H_3SO_3HN : NC_{10}H_6OH$$

then the colour dissolves to a green solution; when in the napthalene group, as in Ponceau SS,

$$C_6H_5N : NC_6H_4N : NC_{10}H_4(SO_3H)_2OH$$

then a violet solution is obtained, and when the HSO_3 groups are contained in both the benzene and napthalene groups, as in the Crocein Scarlets,

$$HSO_3C_6H_4N : NC_6H_4N : NC_{10}H_5OHHSO_3$$

then blue solutions are obtained. These rules, however, do not hold good in all cases; numerous exceptions will be found in the descriptions of the individual colours.

APPLICATION OF THE AZO COLOURING MATTERS.

Cotton Dyeing.—The amido-azo colours (Chrysoidine, Bismarck Brown) are applied on to cotton with a tannin and antimony or tin mordant. They dye easily and well. The primary azo colouring matters are not suitable for cotton dyeing, although they may be applied by first mordanting the cotton in an alumina and oil mordant, but the results are not satisfactory. The secondary azo colours, more especially the Croceines, Biebrich Scarlet, can be dyed by boiling the cotton in a bath containing 5 per cent of alum. Better methods are:—1st. Treat the cotton with stannate of soda 6—8° Tw., steeping for one hour, wring and run through alum; then dye cold or slightly warm. 2nd. Treat the cotton in a bath of oxymuriate of tin 9° Tw. for one to two hours, lift, wring, enter into a bath of acetate of alumina 8° Tw., wring and dye in the cold. (For the tetrazo colours see BENZIDINE COLOUR-ING MATTERS.) *Calico Printing.* — The amido-azo colours are applied with a tannin-thickener, and fixed in a bath of tartar emetic. The acid azo colours are applied with a printing colour made with starch, gum, acetate of alumina and colouring matter; printed, steamed, rinsed and dried. *Wool Dyeing.*—The amido-azo bodies are dyed on wool in a neutral bath containing Glauber's salt. The oxy-azo bodies are dyed in an acid bath containing Glauber's salt and sulphuric or acetic acid. The tetrazo bodies are dyed in a bath containing salt and a trace of acid. *Silk Dyeing.* — The amido-azo bodies are dyed in a bath of soap, old boiled off liquor, or Glauber's salt. The oxy-azo bodies are dyed in a bath of soap, boiled off liquor, or Glauber's salt acidulated with sulphuric acid. The tetrazo dyes in the same way as wool. *Leather* is dyed

by simply brushing over a weak warm solution of the azo dyes.

The affinity of the basic azo colouring matters is so great that it is difficult in a simple bath to ensure level shades, and the same thing happens with some of the oxy-azo colours; the object of adding Glauber's salt is to prevent the too rapid absorption of the colour by the fibre, and so lead to the production of more even dyeings. In the case of the azo colours dyed in an acid bath, the acid is required to liberate the colour-acid from the dye-stuff, and thus to convert the latter into a form in which it can combine with the fibre. *Vide* Dr. Knecht : *Jour. Soc.y. Dyers and Colourists,* 1890. For further details, see METHODS OF APPLYING COAL TAR COLOURS.

Direct production of azo colours on the fibre.—It is essential for the purpose of dyeing by the ordinary means that a colouring matter should be soluble in water, or at least in such a form that will readily combine with any mordant which may be used. Now the simplest azo colours, viz., those produced by azotising the amine and combining with the phenol, are insoluble in water, and in that state are not serviceable for use by dyers. If these colours could be deposited on the fibre they would be very fast, resisting treatment with acids, alkalies, and soaping. The sulphonation of these, while it renders them soluble, has unquestionably a deteriorating influence on their fastness. Several chemists and colour manufacturers, notably Messrs. Read Holliday and Sons, have endeavoured with some success to produce azo colours direct on the fibre, on cotton especially, as on wool and silk the azo colours can be dyed with a fastness that, as a rule, leaves little to be desired. Their first patent is dated 1880, No. 2,757, which has been followed by many others bearing on the same point. The main principle which underlies Holliday's patents is to prepare the cotton with Turkey Red oil, then to pass through a bath of a phenol, such as naphthol, made by solution in caustic soda: wringing and then passing through a final bath containing the amido compound in a diazotised condition, such as would be prepared by dissolving naphthylamine in hydrochloric acid and adding sodium nitrite; or the oiled cotton may be padded in a bath of the amine, diazotised by a passage through a bath of sodium nitrite and acid, then the colour is developed by a passage through an alkaline solution of a phenolic body. By using different amines and phenols various colours can be produced on the fibre, such as yellows oranges, reds, scarlets, etc., and as these are produced in the insoluble condition, the colours are very fast, and many of them of great brilliance; some of the scarlets closely approach Turkey Red in brilliance and in resistance to acids, washing and light.

Dan Dawson produces azo colours on the cotton fibre in a slightly different manner; in his first patent he takes a weak solution of magenta, acidifies with hydrochloric acid, and by adding sodium nitrite azotises it, producing a hexazo body in solution. In this the cotton is entered, worked for a short time, partially dried; then it is passed into a solution of naphthol in caustic soda, when a scarlet is developed on the fibre. In his second patent he takes a weak acid solution of β-naphthylamine in hydrochloric acid, adds sodium nitrite to azotise, neutralises the mixture with chalk, and adds finely precipitated naphthol. Oiled cotton is worked in this mixture until a full scarlet colour is developed.

The same principle has been applied in calico printing by Graessler and Schmidt; a mixture of the amine, naphthol, sodium nitrite, thickening, and acetic acid is printed on the cloth; it is then steamed, which causes the colour to develop. See Schmidt: *Chemiker Zeitung*, 1890. For further details, see METHODS OF APPLYING COAL TAR COLOURS.

Azodiphenyl Blue—
Syn. of induline spirit, soluble.

Azo Acid Rubin 2B—
(DAHL.)—Syn. of amaranth.

Azo Acid Yellow—
(ACTIENGESELLSCHAFT.)—Syn. of Indian Yellow (BAYER).

Azo Eosine—
(BAYER.) 1888.—*Bright scarlet powder,* soluble in water to a scarlet solution; in alcohol and in acetic acid to a red solution, with a bluish red fluorescence; in strong sulphuric acid to a bright crimson solution; on diluting with water a crimson flocculent precipitate is obtained. Hydrochloric acid added to the aqueous solution gives a crimson precipitate. Caustic soda turns the solution yellower.—*Dyes* wool in an acid bath, silk from a broken soap bath fine bright shades of scarlet, fast to acids, light and washing.

Azo Flavin S—
. (BADISCHE.) — Syn. of Indian Yellow (BAYER). .

Azo Fuchsine—
(BAYER.) — Toluene azo - dioxynaphtha-lene sodium monosulphonate

$$(CH_3)C_6H_4N:NC_{10}H_4(OH)_2SO_3Na$$

obtained by diazotising toluidine aᴜ ᴜ combining with dioxynaphthalene sodium monosulphonate. (1890.) *English Patent* 18,517, 1889.—This colouring matter is sent out in two brands, B and G. Azo FUCHSINE B is a dark brownish powder, slightly soluble in cold, easily in hot water to a dark red solution, in alcohol to a scarlet solution, in acetic acid to a magenta coloured solution, in strong sulphuric acid to a crimson solution, from which on diluting with water a flocculent, bright magenta-coloured precipitate falls down. Hydrochloric acid added to the aqueous solution throws down a magenta-red precipitate. Caustic soda turns the colour of the aqueous solution a reddish orange.—*Dyes* from acid baths a deep acid magenta shade, fast to dilute acids, nearly so to strong acids, turned yellower by alkalies, not fast to soaping. Azo FUCHSINE G is a dark violet-red powder, soluble in water to a scarlet solution, in alcohol it is slightly soluble, in acetic acid soluble with a bluish scarlet colour, in strong sulphuric acid to a purple solution, which turns scarlet on diluting with water. Hydrochloric acid and caustic soda have no action on the aqueous solution.—*Dyes* wool and silk from acids baths yellow shades of acid magenta, fast to acids, not fast to soaping, fairly fast to light and stoving.

Azo Green—
(BAYER.)—

$$C-C_6H_4N \begin{cases} C_6H_4N : NC_6H_3OHCO \\ C_6H_4N(CH_3)_2 \\ C_6H_4N(CH_3)_2 \end{cases}$$

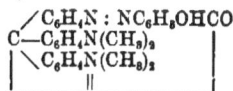

obtained by diazotising meta - amido - tetramethyl - diamido - triphenylmethane combining with salicylic acid, and oxidising the yellow colouring matter to Azo Green. (1890.) *British Patent.—Dark green paste,* insoluble in water, and precipitates slightly on boiling; soluble in alcohol and acetic acid to a green solution, in strong sulphuric acid to an orange-brown solution, which, on diluting with water, turns reddish brown. Hydrochloric acid turns the aqueous solution reddish brown; caustic soda has no action.—*Dyes* wool from an acid bath or on a chrome mordant, bright green; acids turn it yellow; alkalies have no action; fairly fast to light, and moderately so to soaping.

Azoorseillin—

(ACTIENGESELLSCHAFT AND BAYER.)—

Tetrazodiphenyl $\Bigg\langle$ a-naphthol-sodium monosulphonate N.W. / a-naphthol-sodium monosulphonate N.W.

(1883.) *English Patent* 4,237, 1883.—*Dark violet paste*, readily soluble in water to a red-violet solution, in strong sulphuric acid to a blue solution, on diluting with water a violet precipitate forms. Acids do not precipitate the aqueous solution, but redden the colour a little ; caustic soda turns it scarlet.—*Dyes* cotton in an alkaline bath dull shades of claret, not very stable ; exposure to air turns it blue, and it does not dye very even.

Azophosphine—

(LEONHARDT.)—Nitrochrysoidine

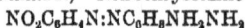

$$NO_2C_6H_4N{:}NC_6H_3NH_2NH_2$$

obtained by diazotising metanitraniline and combining with phenylene diamine. (1886.) *English Patent* 14,962, 1885. Not introduced in this country.

Azorubin S—

(ACTIENGESELLSCHAFT.)—Sodium salt of naphthionic acid azo-a-naphthol monosulphonic acid

$$C_{10}H_6SO_3Na\ N : NC_{10}H_6OHSO_6Na$$

obtained by diazotising naphthionic acid and combining with a-naphthol monosulphonic acid N.W. (1883.) — *Brown powder*, soluble in water to a crimson solution, in strong sulphuric acid to a violet solution, turning crimson on diluting with water. Addition of hydrochloric acid to the aqueous solution produces a brown gelatinous precipitate, while caustic soda turns the solution more scarlet.—*Dyes* wool in acid baths a dark bluish red, very fast to acids, washing and light. Also known as Fast Red C (BADISCHE), Carmoisin (BAYER).

Azorubin 2S—

(ACTIENGESELLSCHAFT.)—Sodium salt of benzene sulphonic acid azo-benzene-azo-a-naphthol sulphonic acid

$$C_6H_4SO_3NaN : NC_6H_4N : NC_{10}H_5OHSO_3Na$$

obtained by diazotising amidoazo-benzene monosulphonic acid and combining with a-naphthol monosulphonic acid N.W. (1883.)—*Brown powder*, soluble in water with a crimson colour, in strong sulphuric acid with a green colour ; on diluting with water this turns first blue then brownish yellow.—*Dyes* wool in acid baths a dark bluish red, fairly fast to acids and washing. This colour is isomeric with Crocein Scarlet 3B (BAYER), and Ponceau 4RB (ACTIENGESELLSCHAFT).

Azo Violet—

(BAYER.)—

Tetrazodiphenetol $\Big\langle$ sodium naphthionate. a-naphthol sodium monosulphonate N.W.

(1886.) *English Patent* 14,424, 1885.— *Brown-black powder*, not very soluble in water to a dark reddish violet solution ; in strong sulphuric acid to a blue solution ; on diluting with water a blue precipitate is obtained. Acids throw down a blue precipitate from the aqueous solution, while caustic soda turns it crimson.—*Dyes* cotton from a soap bath a dark blue violet, moderately fast to light, turned blue by acids, reddened by alkalies. Dyes wool from a salt bath a red shade of violet. Dyes silk reddish violet.

Azo Yellow—

(MEISTER, LUCIUS AND BRUNING.)—Syn. of Indian Yellow (BAYER).

Azulin—

(GUINON, MARNAS AND BONNET.)—Impure hydrochlorate of triphenyl para-rosaniline

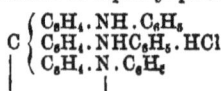

$$C \begin{cases} C_6H_4 \,.\, NH \,.\, C_6H_5 \\ C_6H_4 \,.\, NHC_6H_5 \,.\, HCl \\ C_6H_4 \,.\, N \,.\, C_6H_5 \end{cases}$$

obtained by acting with aniline upon rosolic acid. 1862. Persoz.—*Dark blue powder with bronzy lustre*, insoluble in water, soluble in alcohol to a blue solution, in strong sulphuric acid to a brown-red solution, on diluting with water a blue precipitate of the sulphate is obtained. Hydrochloric acid added to the aqueous solution has little action ; caustic soda changes the colour to a yellowish red. Known also as Azurine.

Azurine—

Syn. of Azuline.

Azurophenoline—

(CHARLES LOWE AND CO.)—*Dark purple grains with a bronzy lustre*, soluble in water and alcohol to a pure blue solution, in strong sulphuric acid to a brownish red solution ; on diluting with water a blue precipitate is obtained. Hydrochloric acid added to the aqueous solution gives a dull blue precipitate ; caustic soda gives a turbid purple solution.—*Dyes* wool and silk from acid baths pure bright blues, fairly fast to light and dilute acids ; strong acids turn it purplish, caustic soda reddens it. Not fast to soaping, which turns the colour a violet-red, while it bleeds strongly.

Anthracene Yellow—

(DAYER.)—Double bromated dioxy-beta-methyl cumarine. (1889.)—*Grey liquid paste*, insoluble in water, soluble in alcohol to a yellow solution, in strong sulphuric acid to olive-brown solution; on diluting with water the original colour is restored. Caustic soda gives an orange - brown solution.—*Dyes* wool mordanted with chrome dull olive-yellows; fairly fast to acids, turned browner by caustic soda, fast to soaping and light. .

ALIZARINE.—To the list of brands of Alizarine given on page 5 should be added the following, manufactured by the British Alizarine Company : Alizarine P, Alizarine YCA, Alizarine SC ; and to the list under Anthropurpurin (page 18) Alizarine SC of the same makers.

B

Basic Blue—

(DURAND AND HUGUENIN.)—Tolyl dimethylamidophenotolylimidonaphthazonium chloride

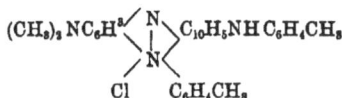

$$(CH_3)_2 NC_6H^2 \diagup^N_N \diagdown C_{10}H_5NH C_6H_4CH_3$$

$$Cl \qquad C_6H_4CH_3$$

obtained by acting with nitrate of nitroso dimethylaniline upon ditolylnaphthylenediamine. (1886.)—*Brown crystalline powder*, soluble in water to a blue-violet solution, in strong sulphuric acid to a greenish brown solution; on diluting with water this turns first green, then violet; finally a blue - violet precipitate is obtained. Hydrochloric acid gives a blue precipitate.—*Dyes* cotton mordanted with tannin and tartar emetic, wool and silk in neutral baths, deep shades of navy blue; fast to acids, light and washing.

Bavarian Blue D.B.F. —

(ACTIENGESELLSCHAFT.) — The sodium salt of diphenylamine blue trisulphonic acid.—*Blue powder*, soluble in water to a blue solution, insoluble in alcohol, soluble in strong sulphuric acid to a yellow-brown solution. On diluting with water this turns first green, then blue. Caustic soda turns the aqueous solution brown - red.—*Dyes* wool and silk in acid baths; cotton requires mordanting with tannin and tartar emetic. The shades given are a dark greenish blue, moderately fast to light and washing.

Bavarian Blue, Spirit Soluble—

Syn. of Diphenylamine Blue.

Bavarian Blue D.S.F.—

(ACTIENGESELLSCHAFT.)—Sodium salt of the di and trisulphonic acids of diphenylamine blue.—*Indigo blue powder*, easily soluble in water to a blue solution, in strong sulphuric acid to a yellow-brown solution; on diluting with water this turns blue. Acids turn the colour of the aqueous solution darker, caustic soda turns it brown-red.—*Dyes* wool and silk in acid baths fine shades of blue.

Bengal Yellow R—

READ HOLLIDAY AND SONS.)—*Brownish yellow powder*, very slightly soluble in cold, soluble in boiling water to a yellow solution, in strong sulphuric acid to a blue solution; on diluting with water a brownish yellow precipitate is obtained. Hydrochloric acid added to the aqueous solution gives a brownish yellow precipitate, caustic soda gives a scarlet solution.—*Dyes* cotton from a soap bath orange-yellow, reddened by soaping, turned purple by acids, scarlet by alkalies.

Bengal Yellow Y—

(READ HOLLIDAY AND SONS.) — *Dark orange powder*, soluble in water to a lemon-yellow solution, in alcohol to a pale yellow solution with green fluorescence, soluble in strong sulphuric acid with a yellowish green fluorescence. Hydrochloric acid added to the aqueous solution turns it orange. Caustic soda decolourises it.—*Dyes* cotton from a salt bath orange-yellow, turned brownish by acids, not fast to soaping or light.

Benzal Green—

Syn. of Malachite Green.

Benzene—

C_6H_6, an aromatic hydrocarbon found present in coal tar. It is the simplest member of the aromatic series of carbon compounds. Molecular weight, 78; specific gravity, 0·8839 ; boiling point, 80° C. Colourless mobile liquid, easily volatile, powerful solvent for all oils and fats; insoluble in water, readily soluble in alcohol, ether, chloroform, glacial acetic acid, being inflammable, burning with a luminous smoky flame. Readily acted on by strong nitric acid, forming nitrobenzene $C_6H_5 NO_2$; strong sulphuric acid, when heated, gradually dissolves benzene, forming benzene-sulphonic acid $C_6H_5 HSO_3$. Not readily acted on by chlorine, especially in the cold or in the dark; in sunlight benzene hexachloride $C_6H_6Cl_6$ is formed. Benzene acts both like a saturated and an unsaturated hydrocarbon; under ordinary circumstances it forms substitution products : nitro-benzene $C_6H_5NO_2$, chlorobenzene C_6H_5Cl, dibromobenzene $C_6H_4Br_2$, etc.; but with chlorine and bromine it forms additive compounds: benzene hexachloride $C_6H_6Cl_6$, benzene hexabromide $C_6H_6Br_6$. According to the views of Kekule, which are generally accepted by most modern chemists, benzene has the constitution shown in the formula

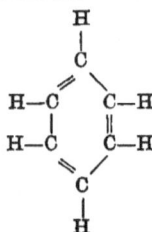

H
|
C
⁄⁄ ＼
H—C   C—H
| ||
H—C   C—H
＼ ⁄⁄
C
|
H

that is the carbon atoms are connected together by three out of the four bonds of affinity they contain in a hexagonal ring, one hydrogen being united with the remaining bond to each carbon atom. This ring of carbon atoms is known as the benzene nucleus, and is present in a large number of the aromatic series of organic compounds. Any view of the constitution of benzene must account for the existence of one monoderivative of the type C_6H_5A, three diderivatives of the type $C_6H_4A_2$, three triderivatives of the type $C_6H_3A_3$, three tetraderivatives of the type $C_6H_2A_4$, one pentaderivative of the type C_6HA_5, and one hexaderivative of the type C_6A_6. The ring formula accounts for these as satisfactorily as any formula can be expected to do. To explain the existence of these isomeric derivatives, the carbon atoms in the benzene ring are numbered thus :—

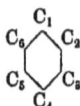

C₁
C₆ ⬡ C₂
C₅   C₃
C₄

All the carbon atoms have an equal value. Therefore there can only be one monoderivative, one pentaderivative, and one hexaderivative, as it is immaterial with what carbon atoms the replacing atom is combined. In the case of the diderivatives the case is different; thus, as the following formula shows, we may have the substituting elements combined with two consecutive carbon atoms (*a*), or these may be combined with two carbon atoms,

separated by one carbon atom (*b*) or two carbon atoms (*c*)

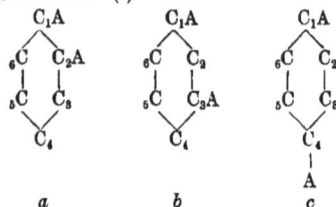

C₁A		C₁A		C₁A		

$$a \qquad b \qquad c$$

In *a* it is immaterial whether the substituting element is with the 2 or the 6 carbon atom, and in *b* whether with the 3 or the 5 atom both give identical compounds. The isomers which have the substituting element in the 1,2 or 1,6 positions are distinguished as *ortho ;* those where the positions are 1,3 or 1,5 as *meta ;* and those where the elements have the position 1,4 as *para.* In the case of the trisubstitution derivatives, the substituting element may have the positions 1,2,3, when it is called *consecutive ;* 1,3,5, when it is called *symmetrical ;* or 1,2,4, when it is called *irregular.* The same system is applied to the tetraderivatives. 1,2,3,4 is *consective*, 1,2,4,5 is *symmetrical*, 1,2,3,5 is *irregular.* For further information as to isomerism other articles on XYLENE, NAPHTHOL, etc., in the present work can be consulted, as also the article BENZENE in the new edition of Watts' "Dictionary of Chemistry."

Benzidine—

$C_6H_4NH_2$
| This base
$C_6H_4NH_2$

has been known for many years, having been discovered by Zinin in 1845 ; but it is only in recent years that it has come into extensive use in colour making. See BENZIDINE COLOURING MATTERS.

Benzidine Blue—

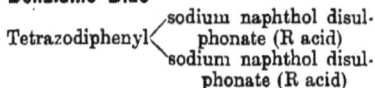

Tetrazodiphenyl ⟨ sodium naphthol disulphonate (R acid)
sodium naphthol disulphonate (R acid)

(1883).—*A coarse powder with a bronzy lustre*, completely soluble in water to a bright blue solution, in strong sulphuric acid to a blue solution ; on diluting with water a violet precipitate is obtained. The aqueous solution turns red on heating, becoming blue again on cooling. Acids produce a red-violet precipitate in the aqueous solution ; caustic soda turns the aqueous solution claret-red.—*Dyes* cotton and wool in neutral baths dirty shades of blue. Not now in use.

Benzidine Colouring Matters—

Under this head it is proposed to describe generally a large number of secondary azo colouring matters derived from benzidin and its homologues. These are also, but somewhat erroneously, termed tetrazo compounds, but they only contain the azo group — N:N— twice and not four times, as the word tetrazo would indicate. However, this term has been generally accepted and will be used here. The colouring matters to be described under this heading are all derived from diamines and their derivatives; mostly from benzidine and its analogues. Benzidine, or para-amidodiphenyl, has the composition shown in the formula—

$$C_6H_4NH_2$$
$$|$$
$$C_6H_4NH_2$$

It is obtained from nitrobenzene by the following series of reactions:—An alcoholic solution of nitrobenzene is treated with caustic soda and zinc dust, whereby it is converted into azobenzene, thus—

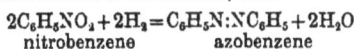

$$2C_6H_5NO_2 + 2H_2 = C_6H_5N:NC_6H_5 + 2H_2O$$
nitrobenzene   azobenzene

This, by boiling with zinc dust, is converted into hydrazobenzene

$$C_6H_5NH - NHC_6H_5$$

by boiling with hydrochloric acid this body undergoes a molecular change, and is converted into benzidine

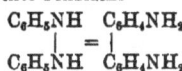

$$\begin{matrix} C_6H_5NH & & C_6H_4NH_2 \\ | & = & | \\ C_6H_5NH & & C_6H_4NH_2 \end{matrix}$$

Benzidine is a crystalline solid, bibasic and forms with hydrochloric acid the chloride

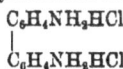

$$C_6H_4NH_2HCl$$
$$|$$
$$C_6H_4NH_2HCl$$

and with sulphuric acid the sulphate

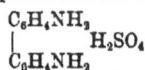

$$\begin{matrix} C_6H_4NH_2 \\ | & H_2SO_4 \\ C_6H_4NH_2 \end{matrix}$$

both of which are used for preparing the colouring matters derived from benzidine. By similar processes, nitrotoluene yields tolidine

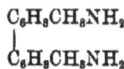

$$C_6H_3CH_3NH_2$$
$$|$$
$$C_6H_3CH_3NH_2$$

nitroxylene yields xylylidin

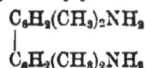

$$C_6H_2(CH_3)_2NH_2$$
$$|$$
$$C_6H_2(CH_3)_2NH_2$$

By first sulphonating paranitrotoluene, and then reducing this body with zinc and

caustic soda, it is converted into diamido-stilbene sulphonic acid

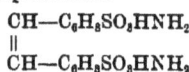

$$CH—C_6H_3SO_3HNH_2$$
$$||$$
$$CH—C_6H_3SO_3HNH_2$$

All these bodies are capable of being diazotised: benzidine to tetrazodiphenyl chloride

$$C_6H_4N:NCl$$
$$|$$
$$C_6H_4N:NCl$$

tolidine to tetrazoditolyl chloride

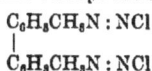

$$C_6H_3CH_3N:NCl$$
$$|$$
$$C_6H_3CH_3N:NCl$$

and the diamidostilbene sulphonic acid to tetrazodiamidostilbene sulphonic acid chloride

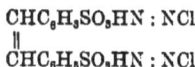

$$CHC_6H_3SO_3HN:NCl$$
$$||$$
$$CHC_6H_3SO_3HN:NCl$$

These diazo compounds can now be conjugated with amines, phenols, or their derivatives, forming colouring matters. Thus, tetrazodiphenyl chloride with naphthionic acid gives the Congo Red

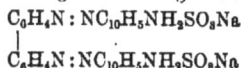

$$C_6H_4N:NC_{10}H_5NH_2SO_3Na$$
$$|$$
$$C_6H_4N:NC_{10}H_5NH_2SO_3Na$$

tetrazoditolyl with salicylic acid gives Chrysamine R

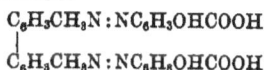

$$C_6H_3CH_3N:NC_6H_3OHCOOH$$
$$|$$
$$C_6H_3CH_3N:NC_6H_3OHCOOH$$

and the stilbene compound with pheno gives Brilliant Yellow

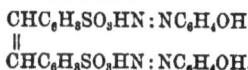

$$CHC_6H_3SO_3HN:NC_6H_4OH$$
$$||$$
$$CHC_6H_3SO_3HN:NC_6H_4OH.$$

It will be seen that one molecule of these diamines is capable of uniting with two molecules of the phenols, amines, etc., to produce the colouring matters. Now it has been noticed (*Eng. Pat.* 15,296, 1885) that this union always takes place in two stages—first one molecule of the amine or phenol combines with the tetrazo body, and then the other; by taking advantage of this fact it is possible to combine one molecule of the tetrazo body with two different phenols, acids, or amines; in this way is produced Congo Corinth, Benzo Orange R, Congo 4R, Congo Yellow, etc. The bases from which these tetrazo colours are made are at present: *p*-amido diphenyl (benzidine) and *p*-amidoditolyl (tolidine) and their derivatives—dianisidin, diphenetidine, etc. *P*-amidostilbene

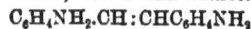

$$C_6H_4NH_2.CH:CHC_6H_4NH_2$$

and its derivatives ;

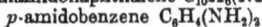

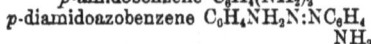

α-diamidonaphthalene $C_{10}H_6(NH_2)_2$
p-amidobenzene $C_6H_4(NH_2)_2$
p-diamidoazobenzene $C_6H_4.NH_2N:NC_6H_4$
 NH_2
diamidocarbazol $C_6H_4NH_2NHNH_2C_6H_4$

The peculiarity which distinguishes these benzidine or tetrazo colours from all others is, that they possess a great affinity for the cotton fibre ; so much so, that they will dye it by simply boiling in a bath containing salt. On what feature in the constitution of these colouring matters this important property depends has not yet been satisfactorily determined ; it is clearly evident that they can be got from a number of diamines ; but what point in the constitution of these diamines gives them this property still requires investigation. It was at first thought that it was because these colouring matters contained a diphenyl group ; but this is now untenable, owing to the existence of colouring matters derived from other diamines. Then it was thought that it was essential that the amidogen groups NH_2 should be in the para position with regard to the other portion of the molecule, but this is not even admissible, as a portion of the *meta* diamines have some affinity for the cotton fibre, so that the feature has still to be found out. In describing these colouring matters it is not proposed to give formulas, owing to their complexity, but to state their composition as far as possible in words, thus :—

Tetrazodiphenyl$\left\{\begin{array}{l}\text{phenol}\\\text{naphthionic acid}\end{array}\right.$

which will, perhaps, simplify matters. Reference may be paid to an article on the benzidin colours in *The Journal of the Society of Dyers and Colourists* for February, 1888, by the author, for further details.

APPLICATION OF THE BENZIDINE COLOURS TO THE FIBRES.

Cotton.—Some of these colours are dyed by boiling in a bath containing soap, potash or soda, and the colouring matters. This method is especially applicable to all the reds of this class. The blues are best dyed in a bath containing soap, Glauber's salt, or borax, and the colouring matters. The yellows are best dyed with soap and phosphate of soda. All of them can be dyed in a bath containing common salt and the colour. *Wool.*—This can be dyed by boiling in a bath containing salt and a little acetic acid. All colours can be used this way, but the results are not all equally satisfactory. *Silk.*—This can be dyed by boiling in a bath con-

taining a little soap and phosphate of soda, or in a salt bath with acetic acid. The reds of this class are not perfectly fast; some, such as Congo, are very fugitive and do not resist acids ; others are a little faster—Erika, Titan Red—to light and acids, but none of them are perfectly satisfactory. It is worth noting that they are decidedly more resistant to light, etc., when dyed on wool than on cotton. The yellows are uniformly fast to light and acids, a few resist boiling in soap, but some are reddened thereby. The blues are not so fast as the yellows, but they are much faster than the reds, and resist acids, but not soaping. For some of this class of colours—those obtained from diamidodiphenyl ether, Benzoazurine 3G, etc.—it has been found that if after dyeing they be passed through a bath of copper or zinc sulphates (*English Patent* 3,934, 1889) they are rendered much faster to light and washing, etc., and in some cases the shade of blue is rendered greener. The colouring matters form an insoluble colour lake, with the metallic base. It is worth noting that fibres dyed with some of these colouring matters, Benzopurpurine, Diamine Red, Diamine Scarlet, etc., lose some of their colour on steeping in cold water for a few hours. The addition of an alkaline salt to the water prevents this loss of colour to some extent. In regard to the question of the use of salt in dyeing with these colouring matters, Gardner (*Textile Manufacturer*, 1890) has shown that some will take a large amount of salt in the bath without any injurious effect ; in other cases there is a maximum limit beyond which the addition of salt retards the dyeing of the colour on to the cloth. In the paper quoted, Gardner gives the best strength of salt solutions to be used with each of the colouring matters of this group. The colouring matters of this group have some affinity for the basic aniline colours (Knecht: *Journal Society Dyers and Colourists*, 1886, page 2), and these may be used to mordant cotton, to be afterwards dyed with the basic anilines ; as a matter of course the shade of the latter is influenced by the particular benzidine colour used. No practical application of this principle is in use among dyers.

Benzidine Red—

(BAYER.)—

Tetrazodiphenyl sodium disulphonate $\left\{\begin{array}{l}\text{sodium naphthionate.}\\\text{sodium naphthionate.}\end{array}\right.$

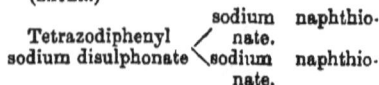

(1883.)—*Brown-red powder*, soluble in water to a crimson-red solution, in strong sulphuric acid to a blue solution, on diluting with water a blue precipitate is obtained. Acids throw down a blue precipitate from the aqueous solution; alkalies do not change it.—*Dyes* cotton in a soap bath reds, which are very weak. Not now in use.

Benzoazurine G—

(BAYER. ACTIENGESELLSCHAFT.) —

Tetrazodi-/ a-naphthol sodium monosulphonate N.W.
phenetol \a-naphthol sodium monosulphonate N.W.

(1885.) *English Patent* 14,424, 1885.— *Blue-black powder*, readily soluble in water to a blue-violet solution, in strong sulphuric acid to a blue solution; on diluting with water a blue-violet precipitate is obtained. Acids give a fine violet precipitate in the aqueous solution. Caustic soda turns it a crimson colour.—*Dyes* cotton from a neutral soap bath dark reddish blue, fast to acids, and fairly fast to light; soaping reddens slightly, and caustic alkalies turn red. Dyes wool from a salt bath dull red-blues. Dyes silk from a neutral soap bath greyish blue shades.

Benzoazurine 3G—

(BAYER.)—

Tetrazodiphenetol< naphthol sodium monosulphonate
\naphthol sodium monosulphonate

(1889.)—*Dark violet-blue powder*, readily soluble in water to a purple solution, slightly soluble to a red-violet solution, in acetic acid to a violet solution, in strong sulphuric acid to a green-blue solution; on diluting with water a violet-red precipitate falls down. Hydrochloric acid added to the aqueous solution throws down a red-violet precipitate; caustic soda turns the colour of the solution to a fine dark crimson. — *Dyes* cotton from a neutral bath, wool from a salt bath, silk from a neutral bath dark blues of a less red tone than Benzoazurine R. Fast to dilute acids, light and washing; caustic soda turns them bright purple.

Benzoazurine R—

(BAYER. ACTIENGESELLSCHAFT.)—

Tetrazodianisol< a-naphthol sodium monosulphonate
\a-naphthol sodium monosulphonate

(1885.) *English Patent* 14,424, 1885.— *Blue-black powder*, readily soluble in water to a dark reddish violet solution, in strong sulphuric acid to a blue solution; on diluting with water a reddish violet precipitate is obtained; acids added to the aqueous solution give a reddish violet precipitate. Caustic soda gives a crimson solution.—*Dyes* cotton from a neutral soap bath dark reddish shades of blue, fast to light and acids; strong soaping slightly reddens it, and alkalies also redden the colour. Dyes wool from a salt bath red shades of dark violet. Dyes silk from a neutral soap bath greyish blue shades.

Benzo Black—

(BAYER.)—1890.—*Greyish black powder*, soluble in water to a dark red-violet solution, in strong sulphuric acid to a dark navy-blue solution; on diluting with water a greenish black precipitate is obtained. Hydrochloric acid added to the aqueous solution gives a dirty blue precipitate; caustic soda has no action.—*Dyes* wool from a salt bath reddish grey to black, turned blue by acids, brown by alkalies, fast to soaping. Dyes cotton from an alkaline bath bluish greys to black, turned blue by acids, violet by alkalies; fast to soaping, fairly fast to light.

Benzo Black S—

Is a stronger brand of this colouring matter.

Benzo Black-Blue G—

(BAYER.)—

Tetrazodiphenyl/ naphthalene azo-naphthol sodium monosulphonate
sodium di-/
sulphonate \ naphthalene azo-naphthol sodium monosulphonate

(1888.) *English Patent.*—*Greyish black powder*, soluble in water to a red-violet solution, insoluble in alcohol, soluble in acetic acid to a dull red-violet solution, in strong sulphuric acid to a brown-red solution; on diluting with water a dirty brownish red precipitate is obtained. Hydrochloric acid added to the aqueous solution gives a violet-blue precipitate; caustic soda has little action.— *Dyes* cotton from a soap bath blackish blue; fairly fast to acids, light, air, and washing.

Benzo Black-Blue R—

(BAYER.)—1888. *English Patent.*— *Dark violet powder*, soluble in water to a red-violet solution, in alcohol to a red-violet solution, in acetic acid to a red-violet solution, in strong sulphuric acid to a dark green - blue solution; on diluting with water a dark violet precipitate is obtained. Hydrochloric acid added to the aqueous solution throws down a red-violet precipitate; caustic soda turns the colour of

D

the aqueous solution redder and brighter.
—*Dyes* cotton from a soap bath dark blue-blacks of a redder tone than the preceding colour. The shades are fairly fast to light, air, acids, and washing.

Benzo Brown B—

(BAYER.) — a-naphthylamine sodium monosulphonate azotriamidoazobenzene hydrochloride

$$C_{10}H_6SO_3NaN : NC_6H_3NH_2N : NC_6H_4NH_2$$
$$HCl$$

(1889.)—*Brown powder*, soluble in water to a reddish brown solution, in alcohol to a dull scarlet solution, in strong sulphuric acid to a navy blue solution ; on diluting with water a brown precipitate is obtained. Hydrochloric acid added to the aqueous solution has little action ; caustic soda gives a yellowish brown precipitate.—*Dyes* cotton from a salt bath reddish brown, turned darker by acids, purplish by alkalies, not fast to soaping, fairly fast to light.

Benzo Brown BX—

Is a stronger brand of this colouring matter.

Benzo Brown G—

(BAYER.)—*Yellow-brown powder*, soluble in water to a reddish brown solution, in alcohol to a brownish orange solution, in strong sulphuric acid to a reddish violet solution; on diluting with water a reddish brown precipitate is obtained. Hydrochloric acid added to the aqueous solution gives a violet-brown precipitate; caustic soda has little action.—*Dyes* cotton, wool, and silk from salt baths reddish browns ; moderately fast to acids, alkalies and light ; fast to soaping.

Benzo Brown NB—

(BAYER.)—1889.—*Brown powder*, soluble in water to a violet-brown solution, in alcohol to a red solution,. in strong sulphuric acid to a dark blue solution ; on diluting with water a purplish brown precipitate is obtained. Hydrochloric acid turns the aqueous solution yellow ; caustic soda gives a pale crimson precipitate in the cold aqueous solution ; on boiling a dirty yellow-brown turbid solution is obtained.—*Dyes* cotton from a salt bath a chestnut-brown, acids turn it purple-brown, alkalies redden it ; not fast to soaping, fairly fast to light.

Benzo Brown NBX—

Is a stronger brand of this colouring matter.

Benzoflavine—

(OEHLER.)—Diamidophenylacridine hydrochlorate, obtained by acting with benzaldehyde upon metatoluylenediamine, treating the product with metatoluylenediamine sulphate, eliminating ammonia by means of hydrochloric acid, and oxidising the product with chloride of iron. (1888.) *English Patent* 9,614, 1888.—*Dull orange powder*, slightly soluble in water to a yellow solution having a faint greenish yellow fluorescence. Soluble in alcohol to a yellow solution with a bright green fluorescence, in acetic acid to a yellow solution with a yellow fluorescence, in strong sulphuric acid to a pale yellow solution with a strong greenish yellow fluorescence ; on diluting with water turns orange-yellow. Hydrochloric acid turns the colour of the aqueous solution darker ; caustic soda almost discharges the colour. —*Dyes* cotton mordanted with tannin and antimony, wool and silk from neutral baths, fine shades of canary yellow; fast to washing ; acids turn it buff.

Benzo Grey—

(BAYER.)—1890.—*Grey powder*, slightly soluble in cold, easily soluble in boiling water, to a reddish brown solution, in strong sulphuric acid to a violet solution ; on diluting with water it turns first reddish brown, then brownish yellow ; a brown precipitate forms. Hydrochloric acid added to the aqueous solution has little action ; caustic soda turns it reddish brown.—*Dyes* wool from a salt bath dull reddish brown, cotton from a soap bath reddish greys, turned blue by acids, purplish by alkalies ; fast to soaping.

Benzo Grey S—

Is a stronger brand of this colouring matter.

Benzo Orange R—

(BAYER.)—

Tetrazodiphenyl{ Alpha sodium naphthionate
sodium salicylate

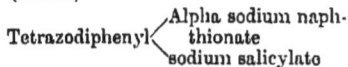

(1890.)—*Scarlet-orange powder*, slightly soluble in cold, easily soluble in hot water, soluble in strong sulphuric acid to a blue solution ; on diluting with water a dark violet-blue precipitate is obtained. Hydrochloric acid added to the aqueous solution gives a dark violet-blue precipitate, acetic acid a blackish brown, and caustic soda an orange precipitate.—*Dyes* cotton from a soap bath reddish orange, not fast to light and air; acids turn it black, caustic soda scarlet, fast to soaping.

Benzopurpurin 1B—

(BAYER. ACTIENGESELLSCHAFT.) —

Tetrazoditolyl $\Big\langle$ B-napthylamine sodium monosulphonate Br. B-napthylamine sodium monosulphonate Br.

(1885.) *English Patent* 3,803, 1885.—
Brown-red powder, not very soluble in water, forming a turbid red solution, soluble in strong sulphuric acid to a blue solution; on diluting with water a blue precipitate is obtained. Hydrochloric acid throws down a blue precipitate from the aqueous solution; caustic soda has no action.—*Dyes* cotton from a soap bath bright scarlets yellower than Congo, not fast to light and dilute acids; fast to soaping and alkalies. Dyes wool from a salt bath duller and yellower shades of red. Dyes silk from a neutral soap bath.

Benzopurpurin 4B—

(BAYER. ACTIENGESELLSCHAFT.) —

Tetrazoditolyl $\Big\langle$ sodium naphthionate sodium naphthionate

(1886.) *English Patent* 3,803, 1885.—
Dark brownish red powder, soluble in water to a turbid red solution, in strong sulphuric acid to a blue solution; on diluting with water a blue precipitate is obtained. Acids added to the aqueous solution throws down the colour acid as a blue precipitate. Caustic soda has no action.—*Dyes* cotton from an alkaline soap bath fine shades of red, slightly yellower than Congo; not fast to light or to acids, which turn it blue; the colour is restored by washing in soap or by alkalies; fast to soaping and alkalies. Dyes wool from a salt bath fine shades. Dyes silk from a neutral soap bath scarlet reds. This colouring matter has been sold under a great variety of names: Royal Scarlet, Cotton Scarlet, Victoria Red 4B, Imperial Red 4B, Eclipse Red 4B, etc.

Benzopurpurin 6B—

(ACTIENGESELLSCHAFT. BAYER.) —

Tetrazoditolyl $\Big\langle$ a-naphthylamine sodium monosulphonate a-naphthylamine sodium monosulphonate Laurents acid

(1885.)—*Red powder*, not very soluble in water to a scarlet solution, in strong sulphuric acid to a blue solution; on diluting with water a blue precipitate is obtained. Acids throw down a blue precipitate from the aqueous solution; caustic soda does not alter it.—*Dyes* cotton from a soap bath bluer shades of red than the 4B, nearly the same shade as Congo; is a little

faster to weak acids and light, but has otherwise the same properties. Dyes wool from a salt bath slightly brighter and paler shades than on cotton. Dyes silk from a neutral soap bath.

Benzopurpurin 10B—

(BAYER.)—1889. *English Patent.*—*Dark brown-red powder*, slightly soluble in cold, easily in hot water, to a dark red solution; soluble in strong sulphuric acid to a dark blue solution, which on diluting with water gives a bright blue precipitate. Hydrochloric acid added to the aqueous solution gives a bright blue precipitate; caustic soda gives a reddish brown precipitate.—*Dyes* cotton from a soap bath deep bluish reds, turned blue by acids, reddened by alkalies, not fast to light.

Benzyl Violet—

Generally the hydrochlorate of pentamethylbenzyl para-rosaniline

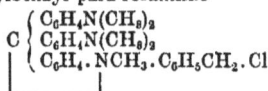

C $\begin{cases} C_6H_4N(CH_3)_2 \\ C_6H_4N(CH_3)_2 \\ C_6H_4 . NCH_3 . C_6H_5CH_2 . Cl \end{cases}$

(1868.) Lauth. By acting with benzyl chloride on methyl violet.—*Brownish metallic lustrous powder* or lumps, with a green glance, soluble in water to a violet solution, in alcohol easily to a deep violet solution, in strong sulphuric acid to a yellow solution; on diluting with water the colour first turns a yellow green, then a greenish blue, finally violet. Hydrochloric acid turns the colour of the aqueous solution first green, then on adding more acid, yellow-brown. Caustic soda gives a brownish yellow precipitate.—*Dyes* wool and silk in neutral baths (cotton requires mordanting with tannin and tartar emetic) violet shades; not very fast to light, acids, and washing. Also known as Paris Violet 6B, Methyl Violet 6B, Violet 5B, Violet 6B.

Biebrich Scarlet—

(KALLE.)—Sodium salt of benzene-sulphonic acid azo-benzene-sulphonic acid azo-B-naphthol.

$C_6H_4SO_3NaN : NC_6H_3SO_3NaN : NC_{10}H_6OH$

obtained by diazotising amidoazobenzene disulphonic acid and combining with B-naphthol. (1878.)—*Red-brown powder*, soluble in water to a scarlet solution, in strong sulphuric acid to a green solution; on diluting with water this turns first blue, then deposits a brown-red flocculent precipitate. Strong aqueous solutions give a brown precipitate on addition of hydrochloric acid; caustic soda gives a brown-red precipitate.—*Dyes* wool and silk in acid baths and cotton in alum baths fine bluish reds, tolerably fast to acids, light and washing. Also known as Imperial Scarlet (BAYER), Ponceau 3RB (ACTIEN-GESELLSCHAFT), Ponceau B (MEISTER, LUCIUS AND BRUNING), Fast Ponceau B (DADISCHE), New Red L (KALLE).

Bismarck Brown—

(ROBERTS, DALE AND CO.)—The hydrochlorate of triamido-azo-benzene

$$H_2C_6H_4N : NC_6H_3(NH_2)_2.2HCl$$

obtained by the action of nitrous acid on *m*-phenylenediamine. (1865.) Introduced by Roberts, Dale and Co. *English Patent* 3,307, 1863.—*Dark brown powder*, soluble in water to a brownish solution, in concentrated sulphuric acid to a brown solution; alkalies precipitate the base from the aqueous solutions, acids cause no change. —*Dyes* wool and silk in neutral bath reddish brown shades; cotton requires a tannin mordant. The shades obtained are tolerably fast to light, air and soaping. Known also as Manchester Brown, Phenylene Brown, Vesuvin, Aniline Brown, Leather Brown, English Brown, Gold Brown, Walnut Brown. The triamido-azo-benzene hydrochloride is distinguished as the red shade of Bismarck Brown (R), while the triamido-azo-toluene hydrochloride, which is produced in a similar way from toluylene diamine, and has similar properties, is distinguished as the Y shade. Bismarck Brown has been combined with other bodies to form direct dyeing browns. *V.* Benzo Brown.

Blackley Blue—

(LEVINSTEIN.)—A soluble induline.

Blue-Black B—

(BADISCHE.)—Sodium salt of *B*-naphthalene sulphonic acid-azo-naphthylamine-azo-*B*-naphthol disulphonic acid

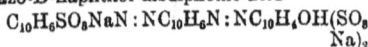

$$C_{10}H_6SO_3NaN : NC_{10}H_6N : NC_{10}H_4OH(SO_3 Na)_2$$

obtained by diazotising *B*-naphthylamine sulphonic acid-azo-*a*-naphthylamine and combining with *B*-naphthol disulphonic acid R. (1882.)—*Blue-black powder*, soluble in water to a dark blue-violet solution, in strong sulphuric acid to a blue-green solution; on diluting with water this turns first blue, then gives a blue precipitate, which is also formed on adding hydrochloric acid to the aqueous solution; caustic soda gives a blue precipitate soluble in water.—*Dyes* wool in acid baths a dark blue-black, the shades being very fast to light, acids and washing. Also known as Azo Black (MEISTER, LUCIUS AND BRUNING).

Bordeaux B—

(BAYER.)—Sodium salt of xylene sulphonic acid azo-xylene sulphonic acid azo-*B*-naphthol

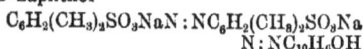

$$C_6H_2(CH_3)_2SO_3NaN : NC_6H_2(CH_3)_2SO_3Na$$
$$N : NC_{10}H_6OH$$

obtained by diazotising amido-azo-xylene disulphonic acid and combining with *B*-naphthol. (1879.) *English Patent* 5,003, 1879.—*Dark blackish red powder*, slightly soluble in cold, more readily in hot water to a dark red solution, easily soluble in alcohol to a deep bluish scarlet solution, in acetic acid to a dark red solution, in strong sulphuric acid to a green solution; on diluting with water a dark red precipitate is obtained. Hydrochloric acid added to the aqueous solution gives a dark red precipitate; caustic soda turns the colour of the solution a little bluer.—*Dyes* wool and silk from acid baths a bluish Bordeaux red, turned darker by acids and alkalies, moderately fast to soaping and light.

Bordeaux S—

(ACTIENGESELLSCHAFT.)—Syn. of maranth.

Bordeaux G—

(BAYER.)—Sodium salt of toluene sulphonic acid azo-toluene sulphonic acid azo-*B*-naphthol sulphonic acid

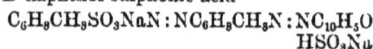

$$C_6H_3CH_3SO_3NaN : NC_6H_3CH_3N : NC_{10}H_5O$$
$$HSO_3Na$$

obtained by diazotising amido-azo-toluene monosulphonic acid and combining with *B*-naphthol monosulphonic acid S. (1879.) *English Patent* 5,003, 1879.—*A dull red powder*, soluble in water to a scarlet-red solution, in alcohol to a red solution, in strong sulphuric acid to a pure blue solution; on diluting with water a dirty red precipitate is obtained. Hydrochloric acid added to the aqueous solution throws down a dark red precipitate. Caustic soda has little action.—*Dyes* wool and silk from acid baths bright Bordeaux red, turned dark purplish red by acids and alkalies, not fast to soaping, moderately fast to light.

Bordeaux Extra—

(BAYER.)—

Tetrazodiphenyl $\Big\langle$
$\begin{array}{l} \text{*B*-naphthol sodium} \\ \text{monosulphonate B.} \\ \text{*B*-naphthol sodium} \\ \text{monosulphonate B.} \end{array}$

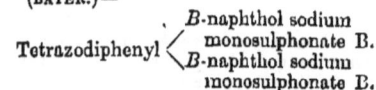

(1883.)—*Brown powder*, soluble in water to a claret-red solution, in strong sulphuric acid to a violet solution, on diluting with water a violet precipitate is obtained.

Hydrochloric acid throws down from the aqueous solution a violet precipitate. Acetic acid does not change the solution; caustic soda turns the aqueous solution yellower.—*Dyes* wool in acid baths claret-reds.

Brahma Orange—

(CARL ZIMMER.)—

Probably

Tetrazodiphenyl $\Big\langle$ sodium diphenylamine monosulphonate sodium salicylate

(1888.) *English Patent* 15,154, 1887.— *Reddish brown powder*, not very soluble in cold, easily in hot water to a brown-yellow solution, partially soluble in alcohol; not soluble in acetic acid, which forms a green turbid mass. Strong sulphuric acid dissolves it with a red colour; on diluting with water a brown flocculent precipitate forms. Hydrochloric acid throws down a greenish precipitate; caustic soda has no action.—*Dyes* cotton from an alkaline soap bath orange-yellow, fairly fast to light and air, reddened by soaping and alkalies, turned green by dilute, blue by strong acids.

Brahma Red B—

(CARL ZIMMER.)—

Probably

Tetrazodiphenyl $\Big\langle$ sodium ethylaniline sulphonate sodium naphthionate

(1888.) *English Patent* 15,154, 1887.— *Dark red powder*, soluble in water to a deep red solution, partially soluble in alcohol, insoluble in acetic acid, which turns the colour to a dark violet; soluble in strong sulphuric acid to a blue solution, from which on diluting with water a reddish brown precipitate falls. Hydrochloric acid throws down a blue precipitate from the aqueous solution; caustic soda has no action.—*Dyes* cotton from an alkaline soap bath fine scarlet, fast to soap and alkali, turned blue by acids, not fast to light.

Brahma Red 6B—

(CARL ZIMMER.)—

Probably

Tetrazoditolyl $\Big\langle$ sodium diphenylamine sulphonate sodium naphthionate

(1888.) *English Patent* 15,154, 1887.— *Dark red powder*, not very soluble in cold, easily in hot water, to a red solution, partially soluble in alcohol. Hydrochloric acid throws down a blue precipitate, acetic acid •a violet precipitate from aqueous solution. Strong sulphuric acid dissolves

it to a blue solution, from which on adding water a brown flocculent precipitate is thrown down; caustic soda has no action. —*Dyes* cotton from an alkaline soap bath bluish reds, not fast to light, acids turn it blue, fast to soap and alkalies.

Brilliant Azurine 5G—

(BAYER.)—

Tetrazodiphenyl-ethylether $\Big\langle$ Dioxynaphthalene sodium monosulphonate Dioxynaphthalene sodium monosulphonate

(1890.)—*Grey powder*, slightly soluble in cold, easily in hot water, to a blue solution, in strong sulphuric acid to a greenish blue solution; on diluting with water a blue precipitate is obtained. Hydrochloric acid added to the aqueous solution gives a blue precipitate; caustic soda turns it crimsor. —*Dyes* cotton from neutral soap baths greenish blue, wool a reddish blue; fairly fast to light, fast to acids, alkalies turn it red, moderately fast to soaping.

Brilliant Black—

(BADISCHE.)—Syn. of Naphthol Black.

Brilliant Congo G—

(ACTIENGESELLSCHAFT. BAYER.)—

Tetrazodiphenyl $\Big\langle$ B-naphthylamine sodium disulphonate B-naphthylamine sodium monosulphonate Br.

(1886.) - *Brick-red crystalline powder*, slightly soluble in water to a brownish red solution, in strong sulphuric acid to a blue solution; on diluting with water a violet precipitate is obtained. Hydrochloric acid added to the aqueous solution gives a dark red-violet precipitate; caustic soda has no action.—*Dyes* cotton in a soap bath a scarlet rather brighter and yellower than Congo, and more resistant to light and acids than that colour. Dyes wool in a salt bath scarlet, the shades being a little duller than on cotton. Dyes silk in a neutral bath bright shades of scarlet.

Brilliant Congo R—

(ACTIENGESELLSCHAFT. BAYER.)—

Tetrazoditolyl $\Big\langle$ B-naphthylamine sodium disulphonate B-naphthylamine sodium monosulphonate Br.

(1886.)—*Brown-red powder*, slightly soluble in water to a red solution, in strong sulphuric acid to a blue solution, on diluting with water a dark brown precipitate is obtained. Hydrochloric acid added to the aqueous solution gives a brown precipitate; caustic soda has no action.—*Dyes* cotton from a soap bath fine bright shades of scarlet a little bluer than Congo but more brilliant; not fast to light, rather faster to acids than Benzopurpurine, which turns them brownish; stands soaping and alkalies. Dyes wool from a salt bath similar shades to those on cotton. Dyes silk from a neutral soap bath fine shades of scarlet.

Brilliant Crocein M—

(CASSELLA AND CO.)—Sodium salt of amidoazobenzene azo-B-naphthol disulphonic acid

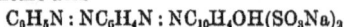

$$C_6H_5N : NC_6H_4N : NC_{10}H_4OH(SO_3Na)_2$$

obtained by diazotising amidoazobenzene and combining with B-naphtholgamma disulphonic acid. (1888).—*Pale brown powder*, soluble in water to a bright red solution, in strong sulphuric acid to a red-violet solution, on diluting with water the solution first turns blue then deposits a brown precipitate. Addition of hydrochloric acid to the aqueous solution gives a brown precipitate; caustic soda turns the colour of the aqueous solution brown.— *Dyes* wool and silk in acid baths fine bright crimson reds, and cotton in a boiling bath with a little alum. Also known as Cotton Scarlet (BADISCHE).

Brilliant Double Scarlet G—

(ACTIENGESELLSCHAFT.)—Sodium salt of B-naphthylamine sulphonic acid azo-B-naphthol

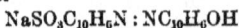

$$NaSO_3C_{10}H_6N : NC_{10}H_6OH$$

obtained by diazotising B-naphthylamine sulphonic acid Br. and combining with B-naphthol.—*Red powder*, soluble in water, in strong sulphuric acid to red solution; on diluting with water a brown precipitate is obtained. Hydrochloric acid added to the aqueous solution gives a brown precipitate.—*Dyes* wool from acid baths bright yellowish scarlet.

Brilliant Double Scarlet 3R—

(BAYER.)—*Dull red powder*, soluble in water to a red solution, in alcohol to a red solution, in acetic acid to a scarlet solution, in strong sulphuric acid to a deep crimson; on diluting with water a red solution is obtained. Hydrochloric acid added to the aqueous solution gives a scarlet precipitate; caustic soda turns it yellower.— *Dyes* wool and silk from acid baths bright scarlets, moderately fast to light, soaping and acids.

Brilliant Green—

(BAYER. BADISCHE. CASSELLA. MEISTER LUCIUS.)— Sulphate or oxalate of tetraethyl di-p-amidotriphenyl carbinol

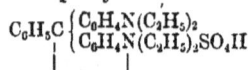

$$C_6H_5C \begin{cases} C_6H_4N(C_2H_5)_2 \\ C_6H_4N(C_2H_5)_2SO_4H \end{cases}$$

obtained by acting with benzaldehyde upon diethylaniline and oxidising the tetraethyldi-p-amidotriphenyl methane produced. (1879). — *Brilliant green crystals* with a golden metallic lustre, soluble in water with a green colour, in strong sulphuric acid to a yellow solution; on diluting with water this turns first reddish yellow, then a yellow-green, finally green. Soluble in alcohol. Hydrochloric acid turns the colour of the aqueous solution reddish yellow; caustic soda precipitates the base as a greenish precipitate.—*Dyes* cotton mordanted with tannin and tartar emetic, wool and silk in neutral baths fine bluish green shades, moderately fast to light, air and washing. Also known as New Victoria Green, Ethyl Green, Smaragd Green, Solid Green J, Fast Green YY.

Brilliant Orange—

(MEISTER, LUCIUS AND BRUNING.)—Syn. of Ponceau 4GB.

Brilliant Ponceau—

(BAYER.) — Syn. of Double Scarlet S Extra.

Brilliant Ponceau—

(CASSELLA.)—Sodium salt of naphthionic acid-azo-B-naphthol disulphonic acid

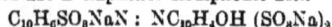

$$C_{10}H_6SO_3NaN : NC_{10}H_4OH (SO_3Na)_2$$

obtained by diazotising naphthionic acid and combining with B-naphthol disulphonic acid G. (1878).—*Scarlet-red powder*, easily soluble in water to a scarlet solution, in strong sulphuric acid to magenta-red solution; on diluting with water this turns scarlet. Hydrochloric acid added to the aqueous solution produces little change; caustic soda turns the aqueous solution brown.—*Dyes* wool in acid baths brilliant scarlet shades, tolerably fast. Also known as New Coccin (ACTIEN.) and Cochineal Red A (BADISCHE). This colour is isomeric with amaranth.

Brilliant Purpurine R—

(ACTIENGESELLSCHAFT. BAYER.)—

$$\text{Tetrazodiphenyl} \begin{cases} B\text{-naphthylamine} \\ \text{sodium disulphonate R} \\ B\text{-naphthylamine} \end{cases}$$

(1888.) *English Patent* 6,087, 1887.— *Dark red powder*, insoluble in cold water, readily soluble in hot water, to a deep red solution. Strong sulphuric acid dissolves it to a dark brown, which does not alter on adding water. Hydrochloric acid turns the colour of the aqueous solution violet; caustic soda has no action.—*Dyes* cotton from a salt bath fine deep bluish red, fast to alkalies and soaping, turned brownish black by acids.

Brilliant Yellow—

(BADISCHE.) — Sodium salt of toluene sulphonic acid azo-diphenylamine

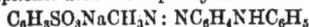

$C_6H_3SO_3NaCH_3N : NC_6H_4NHC_6H_5$

obtained by diazotising toluidine-monosulphonic acid and combining with diphenylamine. — *Orange-yellow powder*, easily soluble in water to a yellow solution, in strong sulphuric acid to a yellow-green solution ; on diluting with water a violet precipitate in a violet solution is obtained. On adding hydrochloric acid to the aqueous solution a violet precipitate is obtained. Caustic soda gives a yellow precipitate. —*Dyes* wool and silk in acid baths a fine orange-yellow.

Brilliant Yellow—

(LEONHARDT. BAYER. ACTIENGESELLSCHAFT.)—

Tetrazostilbene　／Phenol
sodium disulphonate＼Phenol

(1886.) *English Patent 1,387, 1886.*— *Reddish brown powder*, readily soluble in water to a reddish yellow solution, in strong sulphuric acid to a red-violet solution ; on diluting with water a dark violet precipitate is obtained. Hydrochloric acid added to the aqueous solution gives a bluish black precipitate ; caustic soda a scarlet solution.—*Dyes* cotton from a salt bath a very bright orange-yellow, fast to light ; acids turn it blue, soaping orange, alkalies red. Dyes wool from a salt bath similar shades to those on cotton. Dyes silk from a neutral bath orange-yellows.

Brilliant Yellow—

(SCHOLLKOPF ANILINE CO.)—Sodium salt of dinitro-*a*-naphthol monosulphonic acid

$C_{10}H_4(NO_2)_2SO_8NaOH$

obtained by treating *a* - naphthol - disulphonic acid with nitric acid. (1884.)—*Yellow powder*, soluble in water to a brownish yellow solution, turned slightly paler on addition of acids. Caustic soda gives an orange-yellow precipitate.—*Dyes* wool and silk yellow shades in acid bath.

Britannia Violet—

(PERKINS.)—A kind of Hofmans Violet.

Buffalo Rubin—

(SCHOLLKOPF.) - Sodium salt of *a*-naphthylamine-azo-*a*-naphthol disulphonic acid

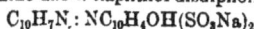

$C_{10}H_7N : NC_{10}H_4OH(SO_3Na)_2$

obtained by treating diazonaphthylamino with *a*-naphthol disulphonic acid Sch. (1884).—*Brown powder*, soluble in water to a magenta-red solution, in strong sulphuric acid to a blue solution, changing to magenta-red on diluting with water ; the aqueous solution does not change on adding hydrochloric acid or caustic soda. —*Dyes* wool or silk in acid bath dark red shades.

Butter Yellow—

Benzene-azo-dimethylaniline

$C_6H_5N : NC_6H_5N(CH_3)_2$

made by combining diazobenzene with dimethylaniline. (1876.) Witt.—*In yellow plates* melting at 115°C., insoluble in water, soluble in dilute hydrochloric acid to a red solution, in strong sulphuric acid to a yellow solution ; on diluting with water a red solution is obtained. Caustic soda gives an orange-yellow precipitate from acid solutions. Is soluble in oils, and is mostly used for colouring butter and fats.

C

Carbazol Yellow—

(BADISCHE.)—

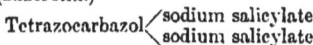

Tetrazocarbazol〈sodium salicylate
　　　　　　　　　＼sodium salicylate

$$HN \left\langle \begin{array}{c} C_6H_3N:N \\ | \\ C_6H_3N:N \end{array} \right.$$

(Tetrazocarbazol)

obtained by azotising carbazoldiamine and combining with salicylic acid. (1889.) *English Patent* 14,478, 14,479, 1888.— *Brownish powder*, slightly soluble in cold, easily in hot water; soluble in sulphuric acid to a dark blue solution ; on diluting with water a violet precipitate is obtained. Hydrochloric acid throws down from the aqueous solution a violet precipitate ; caustic soda has no action.—*Dyes* cotton from neutral soap bath an orange-yellow, fairly fast to soap washing ; acids turn it violet-brown, caustic soda dark red.

Cardinal—

A mixture of Magenta and Chrysoidine.

Cardinal 3B –

(READ HOLLIDAY.)—An azo colour.— *Dark red powder*, readily soluble in water to a bright scarlet solution, in alcohol to a scarlet, in acetic acid to a bluish scarlet solution. Strong sulphuric acid dissolves it with a dark purple colour ; on diluting with water the solution turns scarlet. Hydrochloric acid throws down a streaky crimson precipitate, soluble in water; caustic soda turns the colour of the aqueous solution deeper and yellower.— *Dyes* wool and silk from acid baths a bluish cardinal, fairly fast to acids, washing and light.

Carmoisin -

(BAYER).—Syn. of Azorubin S.

Caroubier —

(DURAND AND HUGUENIN.) — Isomeric with Fast Red E.

Cœrulein—

$$\left[C_6H_4 \left\{ \begin{matrix} CO \cdot C_6H \cdot OH \cdot O \\ C \quad C_6H_2 \cdot O \\ O \end{matrix} \right\rangle O \right.$$

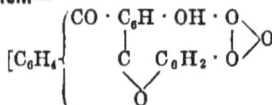

obtained by heating gallein with strong sulphuric acid to 200° C. (1871).—*Black paste*, insoluble in water and alcohol, soluble in strong sulphuric acid to a dirty yellow-brown solution; on diluting with water a reddish brown solution is first obtained, then a greenish solution, then green-black precipitate. Caustic soda added to the paste diluted with water turns it a dirty green colour.—*Dyes* mordanted cotton and wool green shades which are very fast. Also known as Alizarine Green, Anthracene Green.

Cœrulein S—

The bisulphite compound of Cœruleine, made by treating Cœruleine with sodium bisulphite. (1878).—*Black powder*, slightly soluble in cold, easily in boiling water, to a dirty greenish brown solution, soluble in strong sulphuric acid to a dark brown solution; on diluting with water a black precipitate is obtained. Caustic soda turns the colour of the aqueous solution olive-green, and a precipitate is obtained.—*Dyes* mordanted cotton and wool green shades, very fast to light, acids, etc. Known also as Alizarine Green, Anthracene Green.

Canarin —

(DURAND AND HUGUENIN.)—Pseudo-sulphocyanogen

$$SH(CN)_3S_3$$

treating potassium sulphocyanide with oxidising agents. (1849.) Wohler.— *Orange-yellow powder*, insoluble in water, alcohol, or ether; soluble in strong sulphuric acid, reprecipitated by water. Soluble in caustic soda ley, reprecipitated by acids in orange-yellow flocks. Used mostly in calico printing.

Carminaphtha—

(DURAND AND HUGUENIN.)—B-naphthylamine azo-B-naphthol

$$C_{10}H_7N : NC_{10}H_6OH$$

obtained by diazotising B-naphthylamine and combining with B-naphthol. (1880). F)rel.—*Red-brown powder*, insoluble in water, soluble in strong sulphuric acid to a magenta-coloured solution; on adding water the colour is thrown down as a brown-red precipitate. Soluble in alcohol. —*Dyes* wool or silk in acid baths a dull red, which is easily washed out by soap.

Cerise—

Obtained from the crude magenta melt after extracting the magenta.

China Blue —

Syn. of Water Blue, Soluble Blue.

Chrome Violet—

Syn. of Mauveine.

Chrome Violet—

(GEIGY.)—Aurine tricarbonate of ammonia. (1890.) *English Patent* 3,333, 1889.—*Dark purple powder*, easily soluble in boiling water to a dull crimson-red solution, soluble in alcohol to a red solution; insoluble in acetic acid. Strong sulphuric acid gives a brownish yellow solution, turning to a yellow on diluting with water. Hydrochloric acid added to the aqueous solution throws down a brownish red flocculent precipitate in a yellow solution; caustic soda throws down a crimson precipitate.—*Dyes* wool mordanted with chrome a fine reddish violet, fast to dilute acids, darkened by strong acids, reddened by caustic soda; fast to soaping. Very useful for calico printing with a chrome mordant, yielding fine red-violet shades, fast to soaping and light.

Chromotrop 2R—

(MEISTER, LUCIUS AND BRUNING.) — (1891.)—*Dark red powder*, soluble in water, alcohol and acetic acid to scarlet solutions, soluble in strong sulphuric acid to a crimson solution, which turns to a yellowish scarlet on diluting with water. Hydrochloric acid and caustic soda have no action on the aqueous solution.—*Dyes* wool from acid baths a fine bright scarlet, fast to acids; caustic soda turns it brownish. Boiled in soap the colour is discharged. Boiled in bichromate of potassium the colour changes to a violet-black, fast to acids; turned red by caustic soda, fast to soaping. Boiled with solutions of copper, iron, and alumina sulphates, the scarlet colour turns to a dark red, much faster than the original colour.

Chromotrop 2B—

(MEISTER, LUCIUS AND BRUNING.)—1891.
—*Brown-red powder*, soluble in water to a bluish scarlet solution, soluble also in alcohol and acetic acid. Soluble in strong sulphuric acid to a deep crimson solution, which turns scarlet on diluting with water. Caustic soda throws down a red precipitate in a violet fluid from the aqueous solution. Hydrochloric acid has no action.—*Dyes* wool from a strongly acid bath bright red, fast to acids; caustic soda and ammonia turn the shade violet. Not fast to soaping, which discharges most of the colour and turns the shade to a violet. Moderately fast to light. Boiled in a bath of bichromate of potassium, the colour is changed to a black, which is fast to acids and alkalies. On soaping, the colour changes to a bluish tone, but there is no bleeding. Boiled in solutions of copper, iron, or alumina sulphates the scarlet colour is turned dark red.

Chromotrop 6B—

(MEISTER, LUCIUS AND DRUNING.)—1891.
—*Blackish violet powder*, soluble in water, alcohol, and acetic acid to bluish scarlet solutions. In strong sulphuric acid to a deep crimson solution, which on diluting with water gives a turbid scarlet solution. Hydrochloric acid added to the aqueous solution has no action, while caustic soda turns the colour yellower.—*Dyes* wool from strongly acid baths a deep crimson colour, turned bluer by acids, yellower by caustic soda. Boiling in soap discharges and changes the colour to a violet. Boiled in bichromate of potash the colour changes to a greenish black, fast to acids; alkalies turn it bluer, and soaping changes the shade rather bluer, but does not cause it to bleed. Boiled in solutions of copper, iron, or alumina sulphates the crimson colour turns darker and bluer.

Chromotrop 8B—

(MEISTER, LUCIUS AND DRUNING.)—1891.
—*Blackish violet powder*, soluble in water, alcohol, and acetic acid to bluish scarlet solutions, in strong sulphuric acid to a blue solution, which on diluting with water forms a turbid scarlet solution. Hydrochloric acid added to the aqueous solution has no action ; caustic soda darkens the colour.—*Dyes* wool from strongly acid baths a fine bluish crimson; fast to acids, turned purple by caustic soda. Boiled in soap the colour is much reduced and becomes bluer. Boiled with bichromate of potassium the colour changes to a black-blue, which is fast to acids and alkalies. Soaping has little action, the

colour becomes slightly bluer. Boiled with copper, iron, and alumina sulphates the crimson colour changes to a dark red-violet; fast to acids, moderately fast to soaping.

Chromotrop 10B—

(MEISTER, LUCIUS AND DRUNING.)—1891.—*Purple-black powder*, soluble in water, alcohol, and acetic acid to crimson solutions. Soluble in strong sulphuric acid to a blue solution, which does not alter on diluting with water. Hydrochloric acid does not alter the colour of the aqueous solution, while caustic soda turns it rather yellower.—*Dyes* wool from strongly acid baths a fine reddish violet, fast to acids, turned darker by caustic soda. Boiled in soap the colour turns violet and bleeds strongly. Boiled in bichromate of potash the colour changes to a violet-black, fast to acids, reddened by caustic soda. Soaping turns the colour bluer, but does not cause it to bleed. Boiled in solutions of copper, iron, or alumina sulphates the purple colour changes to a dark red-violet, faster to acids and soaping than the original colour.

Chrysamine G—

(BAYER. ACTIENGESELLSCHAFT.)—

Tetrazodiphenyl $\left\langle \begin{array}{l} \text{sodium salicylate} \\ \text{sodium salicylate} \end{array} \right.$

(1884.) *English Patent* 9,162, 1884.—*Yellow-brown powder*, very slightly soluble in cold water, more freely in hot water to a brownish yellow solution, in strong sulphuric acid to a red-violet solution ; on diluting with water a brown precipitate is obtained. Addition of hydrochloric acid to the aqueous solution gives a brown precipitate ; caustic soda turns the colour red.—*Dyes* cotton in a neutral soap bath orange-yellows; fast to light and dilute acids. Strong soaping reddens the shades a little, alkalies turn it red. Dyes wool in a salt bath slightly more orange shades of yellow than on cotton. Dyes silk same as cotton. Was formerly sent out as Flavophenin (BADISCHE).

Chrysamine R—

(BAYER. ACTIENGESELLSCHAFT.)—

Tetrazoditolyl $\left\langle \begin{array}{l} \text{sodium salicylate} \\ \text{sodium salicylate} \end{array} \right.$

(1884.) *English Patent* 9,606, 1884.—*Yellow-brown powder*, not very soluble in water to a brownish yellow turbid solution, in strong sulphuric acid to a red-violet solution ; on diluting with water a brown flocculent precipitate is obtained. Hydrochloric acid added to the aqueous solution gives a brown precipitate ; caustic soda turns the solution red.—*Dyes* cotton from a neutral soap bath orange shades of yellow redder than those from Chrysamine G, but having the same properties. Dyes wool and silk same way as Chrysamine G, but gives redder shades.

Chrysaurein—
Syn. of Orange II.

Chrysaniline—
Syn. of Phosphine.

Chryseolin—
Syn of Resorcin Yellow.

Chrysoidine R.—
(WILLIAMS, THOMAS AND DOWER.)—Introduced in 1876, discovered by Witt. Hydrochlorate of diamidoazobenzene

$$C_6H_5N:NC_6H_3(NH_2)_2HCl$$

obtained by combining diazobenzene with *m*-phenylene diamine. — *In red-brown, lustrous, needle-shaped crystals*, soluble in water to a brownish yellow solution, and in strong sulphuric acid with a brownish yellow solution; on diluting with water this turns red. On adding hydrochloric acid to the aqueous solution it turns red. Caustic soda precipitates the amido base as a red-brown flocculent mass.—*Dyes* wool and silk in neutral baths, cotton with a tannin mordant orange shades, fairly fast to light and air and washing. Has been sold as London Orange. The diamidoazobenzene hydrochloride is distinguished as Chrysoidine R, while the corresponding diamido-azotoluene produced from toluidine and toluylene diamine is known as Chrysoidine Y.

Chrysoin—
Syn. of Resorcin Yellow.

Chrysoline—
(MONNET.) — Sodium salt of benzyl-fluorescein

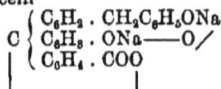

$$C \begin{cases} C_6H_3 . CH_2C_6H_5ONa \\ C_6H_3 . ONa——O \diagup \\ C_6H_4 . COO \end{cases}$$

obtained by heating a mixture of resorcin, phthalic anhydride and benzyl chloride in the presence of sulphuric acid. (1877.)—*Red-brown powder* or lumps smelling of benzyl chloride, easily soluble in water to a brownish solution, which has a strong green fluorescence, in strong sulphuric acid to a yellow solution; on diluting with water a yellow precipitate is obtained. Hydrochloric acid added to the aqueous solution gives a brown-yellow precipitate; caustic soda turns the colour of the aqueous solution darker.—*Dyes* silk from acetic acid

baths yellow, the shades having a weak fluorescence; for wool it is little used; not suitable for cotton.

Chrysophenine C—
(LEONHARDT. BAYER. ACTIENGESELL-SCHAFT.)—

Tetrazostilbene $\diagup$ ethyl phenate sodium disulphonate $\diagdown$ ethyl phenate

got from brilliant yellow by ethylating. (1886.) *English Patent* 4,387, 1886.— *Pale orange powder*, soluble in water to a turbid orange-yellow solution, in strong sulphuric acid to a red-violet solution; on diluting with water a blue precipitate is obtained. Acids added to the aqueous solution give a dark brown-red precipitate, while alkalies do not change it.—*Dyes* cotton from a soap bath fine bright shades of chrome yellow, fast to light, washing and alkalies, turned brown by acids. Dyes wool in a faintly acid bath same shades as on cotton. Dyes silk from a neutral soap or faintly acid bath same shades as on cotton.

Chrysophenine G—
Is the corresponding methyl compound, has similar properties, but gives yellower shades.

Chrysophenol—
$$C_{10}H_{14}N_2O$$
a yellow dye obtained by heating chrys-aniline with hydrochloric acid to 180° F.; forms ruby red needles. Not in use.

Chrysophenoline—
(CHAS. LOWE AND CO.)—1885.—*Brown-ish lustrous grains*, very slightly soluble in cold water, easily soluble in hot water to a reddish orange solution, soluble in strong sulphuric acid to a blue solution; on diluting with water it turns yellowish green. Hydrochloric acid added to the aqueous solution gives a brownish yellow precipitate; caustic soda turns it an orange-red.—*Dyes* wool from slightly acid baths an orange-yellow, silk an olive-yellow, turned darker by acids, red by alkalies, not fast to soap.

Citronine—
(ACTIENGESELLSCHAFT.)—Syn. of Curcumein.

Citronine—
(BROOKE, SIMPSON AND SPILLER.)—A mixture of tetranitrodiphenylamine

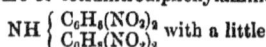

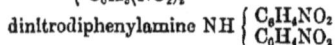

$$NH \begin{cases} C_6H_3(NO_2)_2 \\ C_6H_3(NO_2)_2 \end{cases} \text{with a little}$$

dinitrodiphenylamine $NH \begin{cases} C_6H_4NO_2 \\ C_6H_4NO_2 \end{cases}$

obtained by treating diphenylamine with

nitric acid. 1878. Meldola. *Chem. News,* 1878, vol. 37, p. 56. Witt. *Chem. Socy. Journal,* 1878, p. 210.—*Brownish yellow,* pasty, somewhat crystalline mass, insoluble in water; soluble in alcohol to deep yellow solution, in acetic acid to a lemon-yellow solution, in strong sulphuric acid to a brownish yellow solution, from which on diluting with water a lemon-yellow precipitate is obtained. Caustic soda added to the alcoholic solution turns it to a crimson, and causes it to gelatinise. Hydrochloric acid has no action. Caustic soda added to the paste turns it reddish brown, and forms a flocculent precipitate. —*Dyes* wool and silk from acid baths golden yellow, moderately fast to acids, reddened by alkalies and strong soaping. Not fast to light.

Citronine G—

Syn. of Indian Yellow; a nitro-derivative of Orange IV; also known as Curcumein, New Yellow (which see).

Citronine S—

Syn. of Naphthol Yellow S.

Claret Red S—

Syn. of Amaranth.

Clayton Cloth Red—

(CLAYTON ANILINE CO.)—Ammonia salt of dehydrothioparatoluidine sulphonic acid azo-*B*-naphthol

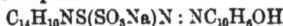

$$C_{14}H_{10}NS(SO_3Na)N : NC_{10}H_6OH$$

(1889.) *English Patent* 14,207, 1889.— *Dark brownish red powder,* soluble in water, alcohol, and acetic acid to an orange-scarlet solution, in strong sulphuric acid to a purple solution; on diluting with water a scarlet precipitate is obtained. Hydrochloric acid added to the aqueous solution gives a scarlet precipitate; caustic soda has no action.—*Dyes* wool from an acid bath or on a chrome mordant dark reds, fairly fast to light, turned brown by acids, not fast to soaping.

Clayton Yellow—

(CLAYTON ANILINE CO.)—Sodium salt of dehydrothioparatoluene sulphonic acid azo-dehydrothioparatoluidine sulphonic acid

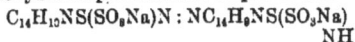

$$C_{14}H_{10}NS(SO_3Na)N : NC_{14}H_9NS(SO_3Na)$$
$$NH_2$$

obtained by diazotising one molecule of the dehydrothioparatoluidine sulphonic acid and combining it with a second molecule of the same body. (1889.) *English Patent* 14,207, 1889.—*Yellow-brown powder,* soluble in water to a yellow solution, in strong sulphuric acid to a yellow-brown solution; on diluting with

water there is no change. Hydrochloric acid added to the aqueous solution gives an orange precipitate; caustic soda a scarlet-orange precipitate.—*Dyes* cotton, wool, and silk from a salt bath chrome yellows, turned orange by acids, scarlet by alkalies, fast to light and soap.

Clematine—

(GEIGY.)—1890.—*Bronzy-brown powder,* soluble in water and alcohol to a crimson solutions, in acetic acid to a violet-red solution, in strong sulphuric acid to a dirty black-violet solution, which, on diluting with water, turns to a dirty violet-brown. Hydrochloric acid added to the aqueous solution turns it blue; caustic soda changes the colour to a bluish red.— *Dyes* wool and silk from a neutral bath, cotton on a tannin mordant, reddish violet shades. Not fast to soaping; acids turn the colour blue, alkalies have no action.

Cloth Brown (Yellow Shade)—

(BAYER.)—

Tetrazodiphenyl $\begin{cases} \text{sodium salicylate} \\ \text{dioxynaphthaline} \end{cases}$

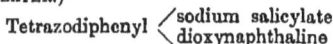

(1889).—*Black powder,* soluble in water to a dark-brown solution, in strong sulphuric acid to a reddish violet solution; on diluting with water a brown precipitate is obtained. Caustic soda has no action on the aqueous solution.—*Dyes* wool from acid baths, or on a chrome mordant, dark red brown, fast to acids and washing and light.

Cloth Brown (Red Shade)—

(BAYER.)—

Tetrazodiphenyl $\begin{cases} \text{sodium salicylate} \\ \text{naphthol sodium} \\ \text{sulphonate} \end{cases}$

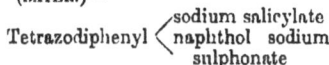

(1889).—*Claret-brown powder,* soluble in water to an orange-red solution, in strong sulphuric acid to a reddish violet solution; on diluting with water a scarlet precipitate is obtained. Caustic soda has no action on the aqueous solution.—*Dyes* wool from an acid bath, or on a chrome mordant, dark brownish red shades, fairly fast to light and soaping, reddened by acids.

Cloth Orange—

(BAYER.)—

Tetrazodiphenyl $\begin{cases} \text{sodium salicylate} \\ \text{dioxybenzene} \end{cases}$

(1889.)—*Light chocolate powder,* soluble in water to an orange solution, in strong sulphuric acid to a reddish violet solution; on diluting with water an orange precipitate is obtained. Hydrochloric acid added to the aqueous solution gives an orange precipitate; caustic soda has no action.—*Dyes* wool from acid bath or on a chrome mordant bright orange; fast to acids and light, not fast to soaping.

Cloth Red B—

(OEHLER. BAYER AND CO.)—Sodium salt of amido-azo-toluene-azo-B-naphthol di-sulphonic acid

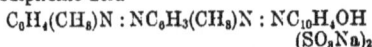

$$C_6H_4(CH_3)N : NC_6H_3(CH_3)N : NC_{10}H_4OH$$
$$(SO_3Na)_2$$

by diazotising amido-azo-toluene and combining with B-naphthol disulphonic acid R. (1879.) *English Patent* 5,003, 1879. —*Dark brown powder*, soluble in water to a dark bluish red solution, in strong sulphuric acid to a blue solution; on diluting with water a brown-red precipitate is formed. Hydrochloric acid added to the aqueous solution turns it brown.— *Dyes* wool in acid baths a dark red of a brownish tone; very fast to light, acids, and washing.

Cloth Red 3B Extra—

(BAYER.)—*Dark red powder*, soluble in hot water to a bluish red solution, slightly soluble in alcohol, or in acetic acid; soluble in strong sulphuric acid to a dark-red solution, which, on diluting with water, turns brown. Hydrochloric acid added to the aqueous solution makes it darker, while caustic soda turns it bluer. —*Dyes* wool in acid baths bright bluish reds, fast to acids, light, and washing.

Cloth Red G—

(OEHLER. BAYER AND CO.)—Sodium salt of amido - azo - toluene - azo- B - naphthol monosulphonic acid

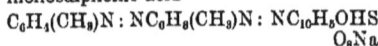

$$C_6H_3(CH_3)N : NC_6H_6(CH_3)N : NC_{10}H_5OHS$$
$$O_3Na$$

obtained by diazotising amido-azo-toluene and combining with B-naphthol monosulphonic acid S. (1879.) *English Patent,* 5,003, 1879.—*Red-brown powder*, slightly soluble in water to a brownish red solution, and in strong sulphuric acid to a blue solution; on diluting with water a brown-red precipitate forms. On adding hydrochloric acid to the aqueous solution the same precipitate is obtained.—*Dyes* wool in acid baths a dark red, very fast to light, acids, and washing.

Cloth Red G Extra—

(BAYER.) — Benzene-azo-benzene azo-naphthol sodium monosulphonate

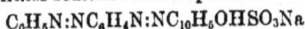

$$C_6H_5N:NC_6H_4N:NC_{10}H_5OHSO_3Na$$

(1880.)—*English Patent* 5,003, 1879.— *Light red powder*, soluble in water to a deep red solution, slightly soluble in

alcohol; soluble in strong sulphuric acid to a blue solution, on diluting which with water a brown-yellow precipitate forms. Hydrochloric acid added to the aqueous solution turns it yellowish red; caustic soda has no action.—*Dyes* wool and silk from acid baths fine reds, fast to acids, light, and washing.

Coccinin—

Syn. of Phenetol Red.

Coccinin B—

(MEISTER, LUCIUS AND BRUNING.)— Sodium salt of p-cresolmethylether-azo-B-naphthol disulphonic acid

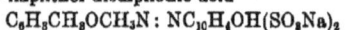

$$C_6H_3CH_3OCH_3N : NC_{10}H_4OH(SO_3Na)_2$$

obtained by diazotising amido-p-cresol-methylether and combining with B-naph-thol disulphonic acid R. (1878.)—*Dark red powder*, soluble in water to a cherry-red solution, in strong sulphuric acid to a cherry-red solution, which is not changed on diluting with water; acids added to the aqueous solution darken it, while caustic soda gives a brown precipitate in a red-brown solution.—*Dyes* wool and silk red shades in acid baths, not very fast.

Cochineal Red—

(CASSELLA.)—Syn. of Brilliant Ponceau.

Cochineal Scarlet G—

(SCHOLLKOPF COLOUR CO.)—Sodium salt of benzene-azo-a-naphthol monosulphonic acid C.

$$C_6H_5N : NC_{10}H_5OHSO_3Na$$

obtained by combining diazobenzene with a-naphthol monosulphonic acid C. (1883.) *English Patent* 15,781, 1885.—*Brick-red powder*, soluble in water with a yellowish red colour, in strong sulphuric acid to a cherry-red solution; on diluting with water a brown-red precipitate falls down. Hydrochloric acid gives a precipitate of the free colour acid; caustic soda produces an orange-yellow coloured solution. — *Dyes* wool in acid bath dull reds.

Cochineal Scarlet 2R—

(SCHOLLKOPF.)—Sodium salt of toluene-azo-a-naphthol monosulphonic acid

$$C_6H_4CH_3N : NC_{10}H_5OHSO_3Na$$

obtained by treating diazotoluene with a-naphthol monosulphonic acid C. (1883.) *English Patent* 15,781, 1885.—*Vermilion-red powder*, difficultly soluble in cold, easily soluble in hot water to a yellowish red solution, in strong sulphuric acid to a magenta-red solution; on diluting with water a flocculent red precipitate is obtained. Hydrochloric acid added to the aqueous solution precipitates the colour acid in red flocks; caustic soda changes the colour of the aqueous solution to orange.—*Dyes* wool and silk in acid baths bright shades of red closely resembling those obtained from cochineal.

Cochineal Scarlet 4R

(SCHOLLKOPF.)—Sodium salt of xylene-azo-*a*-naphthol monosulphonic acid

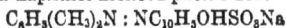

$$C_6H_3(CH_3)_2N : NC_{10}H_5OHSO_3Na$$

obtained by treating diazoxylene with *a*-naphthol monosulphonic acid C. (1883.) *English Patent* 15,781, 1885.—*Fiery red powder*, slightly soluble in water, soluble in strong sulphuric acid to a magenta-red solution; on diluting with water a reddish precipitate is obtained; on adding hydrochloric acid to the aqueous solution a red-brown precipitate is obtained; the colour of the aqueous solution is turned orange by caustic soda.—*Dyes* wool and silk in acid baths bright reds, fast to light, air, and soaping.

Congo—

(ACTIENGESELLSCHAFT. BAYER).—

Tetrazodiphenyl $\langle$ sodium naphthionate / sodium naphthionate

(1884.) BOTTIGER. *English Patent* 4,415, 1884.—*Dark red powder*, soluble in water to a dark red solution, in strong sulphuric acid to a blue solution; on diluting with water a blue precipitate is obtained. Hydrochloric acid gives a blue precipitate in the aqueous solution; acetic acid gives a blue-violet precipitate; a strong solution of caustic soda gives in strong solutions of the colour a red precipitate soluble in water.—*Dyes* cotton in a soap bath bright scarlet-red shades, not permanent, becomes dark on exposure to air, and fades on exposure to light. Acids turn them blue; fast to soaping and washing. Dyes wool in neutral bath dull shades of red.

Congo GR—

(ACTIENGESELLSCHAFT. BAYER.)—

Tetrazodi- / sodium *m*-nitrosalicylate phenyl \ sodium naphthionate

(1885.) *English Patent. — Dark red powder*, freely soluble in water to a dark red solution, in strong sulphuric acid to a blue solution; on diluting with water a blue precipitate is obtained. Acids throw down a blue precipitate from the aqueous solution; alkalies brighten the colour a little.—*Dyes* cotton in a soap bath bright scarlet, not fast; acids turn it blue, stands soaping. Dyes wool in neutral bath bright reddish orange shades.

Congo 4R—

(ACTIENGESELLSCHAFT. BAYER.)—

Tetrazoditolyl $\langle$ sodium naphthionate / resorcin

(1885.) *English Patent* 4,415, 1884.—*Dark red powder*, slightly soluble in water to a dark reddish solution, in strong sulphuric acid to a blue solution; on diluting with water a violet precipitate is obtained. Hydrochloric acid gives a dark purple-red precipitate in the aqueous solution; caustic soda does not alter it.—*Dyes* cotton from a soap bath bluish shades of scarlet, rather more stable than Congo or Benzopurpurin; acids dull the colour considerably; fast to soaping. Dyes wool from a salt bath brighter shades than on cotton, and which are more stable. Dyes silk from a neutral soap bath shades like those got on cotton.

Congo Brown G—

(ACTIENGESELLSCHAFT.)—1890. *English Patent* 10,653, 1888.—*Grey powder*, slightly soluble in cold, easily soluble in hot water to a brownish orange solution, soluble in alcohol to a yellow solution, in strong sulphuric acid to a dark red solution, on diluting which with water a flocculent reddish orange precipitate falls down. Hydrochloric acid added to the aqueous solution gives a reddish brown precipitate; caustic soda changes the colour of the solution to scarlet.—*Dyes* cotton from a salt or neutral bath brownish orange shades; turned dark blue by strong acids, purple-brown by dilute acids, bluish red by caustic soda; not fast to soaping or light. Dyes wool from salt bath brownish orange shades as with cotton.

Congo Brown R—

(ACTIENGESELLSCHAFT.)—1890. *English Patent* 10,653, 1888.—*Grey powder*, soluble in water to a dull red solution, in alcohol to an amber-coloured solution, in acetic acid to a pale red solution, in strong sulphuric acid to a dark crimson solution, on diluting which with water a flocculent orange precipitate is obtained. Hydrochloric acid throws down a red precipitate from the aqueous solution, while caustic soda has no action.—*Dyes* cotton, wool, and silk from a salt or neutral bath orange-brown, turned purplish brown by acids, bluish red by alkalies; not quite fast to soaping.

Congo Corinth—

(ACTIENGESELLSCHAFT. BAYER.)—

Tetrazodiphenyl $\langle$ *a*-naphthylamine sodium monosulphonate / *a*-naphthol sodium monosulphonate N.W.

(1886). *English Patent* 15,296, 1885; 2,213, 1886.— *Dark greenish powder,* readily soluble in water to a dark crimson-red solution, in strong sulphuric acid to a blue solution; on diluting with water a blue-violet precipitate is obtained. Hydrochloric acid throws down the free colour acid as a blue precipitate; caustic soda turns the colour of the aqueous solution crimson.—*Dyes* cotton in soap bath a purplish red, not fast; acids turn it blue, alkalies turn it red. Dyes wool a dull reddish scarlet.

Congo Corinth B—

(ACTIENGESELLSCHAFT. BAYER.)—

Tetrazoditolyl $\Big\langle$ sodium naphthionate $\atop$ a-naphthol sodium sulphonate N.W.

(1885). *English Patent* 15,296, 1885; 2,213, 1886.—*Brownish black powder,* not readily soluble in water to a dark crimson solution, in strong sulphuric acid to a violet solution; on diluting with water a violet precipitate is obtained. Hydrochloric acid added to the aqueous solution throws down a blue precipitate; caustic soda turns the colour of the solution dark crimson.—*Dyes* cotton from a soap bath a dark reddish violet, moderately fast to light, turned blue by acids, fast to soaping, reddened by alkalies. Dyes wool plum colours.

Congo Fast Blue B—

(ACTIENGESELLSCHAFT.) — 1890. — *Dark brown powder,* soluble in water to a blue solution, very slightly soluble in alcohol, soluble in acetic acid. Strong sulphuric acid gives an olive-brown solution, which, on diluting with water, turns to a turbid brownish yellow solution. Hydrochloric acid added to the aqueous solution gives a blue precipitate; caustic soda has no action. —*Dyes* cotton from a neutral soap bath pale bright blue; turned greenish by strong acids, violet-blue by caustic soda; fast to soaping. Dyes wool from a neutral bath deep red-blue, not fast to soaping.

Congo Fast Blue R—

(ACTIENGESELLSCHAFT.)—*Brown powder,* soluble in water to a reddish blue solution, only slightly soluble in alcohol, soluble in acetic acid to a dark violet solution, in strong sulphuric acid to a greenish brown solution, which on diluting with water turns greenish yellow. Hydrochloric acid throws down a reddish blue precipitate, and caustic soda a red-violet precipitate from the aqueous solution.—*Dyes* cotton from a neutral soap bath pale bright blue, reddened by acids, turned blue by caustic

soda; not quite fast to soaping. Dyes wool from neutral baths deep red-blues, not fast to soaping.

Congo Orange R—

(ACTIENGESELLSCHAFT.)—

Tetrazodiphenyl $\Big\langle$ Beta-naphthylamine $\atop$ sodium disulphonate R. $\atop$ ethyl phenol ether $(C_6H_4OC_2H_5)$

(1890.)—*Scarlet powder,* soluble in water and alcohol to an amber solution, in strong sulphuric acid to a deep indigo solution; on diluting with water a light purple - brown precipitate is obtained. Hydrochloric acid added to the aqueous solution gives a purple-brown precipitate; caustic soda brightens the solution.—*Dyes* cotton from a soap bath reddish orange, turned purple by acids, redder by alkalies; fast to soaping, not fast to light.

Congo Yellow—

(ACTIENGESELLSCHAFT. BAYER.)—

Tetrazodiphenyl $\Big\langle$ sodium sulphanilate $\atop$ phenol

(1885.)—*Brown-yellow paste,* only slightly soluble in water, in strong sulphuric acid to a brownish red solution; on diluting with water a brown precipitate is obtained. Acids give a brown-yellow precipitate in the aqueous solution; caustic soda gives a red solution.—*Dyes* cotton and wool in neutral baths yellows of a slightly orange tone. These are very sensitive to alkalies or soap, which redden them, or to acids, which turn them a dirty greenish yellow.

Coralline—

Sodium salt of Aurine; obtained by treating Aurine with soda. — *Resinous brittle mass* with a green metallic lustre, soluble in water to a red solution, in alcohol to a magenta solution, in strong sulphuric acid to a yellow solution; on diluting with water a yellow precipitate is obtained. Hydrochloric acid changes the colour of the aqueous solution to yellow and gives a brownish yellow precipitate; caustic soda causes little change. Used for making lake colours for calico-printing.—*Dyes* wool and silk from very faintly acid solutions bright reds, not fast to acids, etc. Known also as Yellow Coralline, Soluble Aurine.

Cotton Blue—

Syn. of Water Blue.

Cotton Blue R—

(BADISCHE.)—Syn. of Meldolas Blue.

Cotton Brown A—

(CASSELLA.) — 1889. — *Blackish brown powder*, not readily soluble in cold water, more easily in hot, to a reddish brown solution ; in alcohol it easily dissolves to an orange-brown solution, in acetic acid it is only slightly soluble; in strong sulphuric acid it dissolves with a blue-violet colour; on diluting the solution with water a dark brown precipitate is thrown down. Hydrochloric acid added to the aqueous solution throws down a dark brown precipitate ; caustic soda turns the colour to a deep red. —*Dyes* cotton from a soap bath reddish brown, fast to soap and washing, turned darker by acids.

Cotton Brown N—

(CASSELLA.) — 1889. — *Brown powder*, soluble in water to a reddish brown solution, in strong sulphuric acid to a brown-yellow solution, which turns brown on diluting with water. Hydrochloric acid darkens the colour of the aqueous solution; caustic soda has no action.— *Dyes* cotton from a salt bath reddish browns, fast to washing, acids, and fairly fast to light.

Cotton Scarlet—

(BADISCHE.)—Syn. of Brilliant Croceine.
(CASSELLA.)

Cotton Scarlet—

A mixture of safranine and chrysoidine.

Cotton Violet—

Liquid made from magenta by treatment with crude wood spirit, now not used.

Cotton Yellow G -

(BADISCHE.)—

$$CO\left\langle{\begin{array}{l} NHC_6H_4N : NC_6H_3OHCOONa \\ NHC_6H_4N : NC_6H_3OHCOONa \end{array}}\right.$$

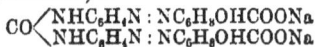

obtained by acting with phosgene on amido-benzene-azo-salicylic acid. (1889.) *English Patent* 15,258, 1888.—*Brownish yellow powder*, soluble in water to a brownish yellow solution, in sulphuric acid to a scarlet solution ; on diluting with water a dark brownish red precipitate is obtained. Hydrochloric acid added to the aqueous solution produces a brownish red precipitate ; caustic soda causes little change.—*Dyes* cotton from a neutral soap bath bright yellow, fairly fast to light and soap, turned red by alkalies, brown by acids.

Couplers Blue—

Syn. of Induline spirit soluble.

Cresol Red—

(BADISCHE.)—Sodium salt of o-cresol-

ethylether - azo - β-naphthol disulphonic acid

$$C_6H_3CH_3OC_2H_5N : NC_{10}H_4OH(SO_3Na)_2$$

obtained by diazotising amido-o-cresolethylether and combining with β-naphthol disulphonic acid R. (1878.) Properties similar to those of Coccinin B, but dyes bluer shades of red.

Cresotine Yellow G—

(OEHLER.)—

$$Tetrazodiphenyl\left\langle{\begin{array}{l} o\text{-cresolcarboxylic acid} \\ o\text{-cresolcarboxylic acid} \end{array}}\right.$$

(1889.) *English Patent* 7,997, 1888.— *Pale ochre brown powder*, slightly soluble in cold, easily in hot water to a brownish yellow solution ; in strong sulphuric acid to a deep purple solution, which turns brown on diluting with water. Hydrochloric acid added to the aqueous solution produces an olive-brown precipitate; caustic soda gives a rather turbid red solution.— *Dyes* cotton from a neutral soap bath, or wool and silk from neutral baths, sulphur yellows, moderately fast to light, turned brown by acids, red by alkalies, orange by soaping.

Cresotine Yellow R—

(OEHLER.)—

$$Tetrazoditolyl\left\langle{\begin{array}{l} o\text{-cresolcarboxylic acid} \\ o\text{-cresolcarboxylic acid} \end{array}}\right.$$

(1889.) *English Patent* 7,997, 1888.— *Reddish brown powder*, slightly soluble in cold, easily soluble in hot water to a reddish yellow solution, in strong sulphuric acid to a bright purple solution; on diluting with water a purplish precipitate is obtained. Hydrochloric acid gives a purplish precipitate in the aqueous solution ; caustic soda gives an orange-red solution.—*Dyes* cotton from a neutral soap bath, wool and silk from neutral baths, orange-yellows, turned purple-brown by acids, red by alkalies.

Crimson—

Is a red colouring matter obtained during the purification of the crude magenta, and has similar properties, but dyes bluer shades.

Crocein AZ—

(CASSELLA AND CO.)—1891.—*Brownish red powder*, soluble in water to a crimson solution, in alcohol to a scarlet solution, in acetic acid to a crimson solution, in strong sulphuric acid to a blue solution; from which on adding water a reddish orange flocculent precipitate is thrown down. Hydrochloric acid turns the colour of the aqueous solution brown, caustic soda a purplish red.—*Dyes* cotton from a boiling bath containing alum bright bluish pink, turned blue by strong acids, violet by caustic soda ; not quite fast to soaping. Dyes wool from neutral· baths deep bluish scarlet, not fast to soaping.

Crocein B—

(SCHOLLKOPF.)—Sodium salt of amido-azo - benzene - azo - a - naphthol disulphonic acid

$C_6H_5N : NC_6H_4N : NC_{10}H_4OH(SO_3Na)_2$

obtained by diazotising amido-azo-benzol and combining with a-naphthol disulphonic acid Sch. (1884.) *English Patent* 15,775, 1885.—*Brown-red powder*, slightly soluble in water to a magenta-red solution, in strong sulphuric acid to a violet solution ; on adding water a red-violet precipitate falls down ; on adding hydrochloric acid to the aqueous solution a brown flocculent precipitate forms, which is soluble on adding more water. Caustic soda turns the colour of the aqueous solution violet.—*Dyes* wool and silk in acid baths a fine bluish red, tolerably fast, and will dye cotton a red that, however, does not resist washing.

Crocein 3B—

(SCHOLLKOPF.)—Sodium salt of amido-azo - toluene - azo - a - naphthol disulphonic acid

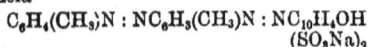

$C_6H_4(CH_3)N : NC_6H_3(CH_3)N : NC_{10}H_4OH$
$(SO_3Na)_2$

obtained by diazotising amido-azo-toluene and combining with a-naphthol disulphonic acid Sch. (1884.)—*Dark brown powder*, soluble in water to a crimson or magenta-red solution, in strong sulphuric acid to a blue solution ; on diluting with water a violet precipitate first forms, which re-dissolves on adding more water, a crimson solution being obtained. On adding hydrochloric acid to the aqueous solution a violet precipitate is thrown down. Caustic soda turns the aqueous solution brown.—*Dyes* wool and silk crimson shades in acid baths, and cotton in a boiling bath. The shades are pretty fast to light and washing.

Crocein 3BX —

(BAYER.)—Sodium salt of naphthionic acid-azo-B-naphthol monosulphonic acid

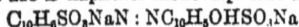

$C_{10}H_6SO_3NaN : NC_{10}H_5OHSO_3Na$

obtained by diazotising naphthionic acid and combining with B-naphthol monosulphonic acid B. (1882.) *English Patent* 2,080, 1881.—*Scarlet-red powder*, soluble in water to a red solution, in strong sulphuric acid to a red-violet solution ; on diluting with water a red solution is obtained. Addition of hydrochloric acid to the aqueous solution produces no change,

while caustic soda turns the colour yellow-brown.—*Dyes* wool in acid baths a bright red. Isomeric with Fast Red E, Azo Rubin S.

Crocein Orange—

(KALLE. BAYER AND CO.)—Syn. of Ponceau 4GB.

Crocein Scarlet 3B—

(BAYER).—Sodium salt of benzene sulphonic acid azo-benzene-azo-B-naphthol sulphonic acid

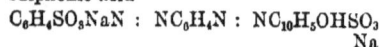

$C_6H_4SO_3NaN : NC_6H_4N : NC_{10}H_5OHSO_3$
Na

obtained by diazotising amido-azo-benzene monosulphonic acid, and combining with B-naphthol monosulphonic acid B. (1881.) *English Patent* 1,225, 1881.—*Red-brown powder*, soluble in water to a scarlet solution, in strong sulphuric acid to a pure blue solution ; on diluting this with water a yellow-brown precipitate falls, which re-dissolves on adding more water to a red solution. On adding hydrochloric acid to the aqueous solution a yellow-brown flocculent precipitate falls. Caustic soda turns a weak aqueous solution a dirty violet-red colour ; in strong solutions a violet-red precipitate falls.—*Dyes* wool and silk in acid baths bluish reds, and cotton in an alum bath, the shades are fast to light, soap, and acids. New Red 5R (BAYER) is isomeric with this colouring matter ; also known as Ponceau 4RB (ACTIENGESELL-SCHAFT).

Crocein Scarlet 7B—

(BAYER.)—Sodium salt of toluene sulphonic acid azo-toluene-azo-B-naphthol sulphonic acid

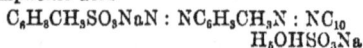

$C_6H_3CH_3SO_3NaN : NC_6H_3CH_3N : NC_{10}$
H_5OHSO_3Na

obtained by diazotising amido-azo-toluene monosulphonic acid, and combining with B-naphthol monosulphonic acid B. (1881.) *English Patent* 1,225, 1881.—*Red brown powder*, soluble in water to a scarlet solution, in strong sulphuric acid to a blue solution ; on diluting with water this turns violet-red. Addition of hydrochloric acid to the aqueous solution turns it to a rather turbid crimson solution ; caustic soda turns it to a dirty violet colour.—*Dyes* wool in acid baths, cotton in alum baths, fine bluish reds. Also known as Ponceau 6RB (ACTIENGESELLSCHAFT). This colour is isomeric with Orseillin BB (BAYER) and Bordeaux G (BAYER).

Crocein Yellow—

(BAYER.)—This was obtained by treating *Beta*-naphthol monosulphonic acid R with nitric acid. It is not now used.

Crystal Scarlet B—

(DADISCHE.)—Syn. of Crystal Scarlet 6R.

Crystal Scarlet R—

(DADISCHE.)—Syn. of Crystal Scarlet 6R (CASSELLA).

Crystal Scarlet 6R—

(CASSELLA AND CO.)—Sodium salt of a-naphthylamine - azo - B - naphthol disulphonic acid

$$C_{10}H_7N : NC_{10}H_4OH(SO_3Na)_2$$

obtained by treating diazonaphthylamine with B-naphthol disulphonic acid G (*gamma*). (1885.) *English Patent* 816, 1884.—*Beautiful brownish red crystals with a golden lustre,* soluble in water to a scarlet solution, and in strong sulphuric acid to a violet solution, which on diluting with water turns to a scarlet colour. On adding a little hydrochloric acid the aqueous solution turns a little darker in colour; on adding much acid the free colour acid is precipitated in the form of fine brown crystalline plates. Caustic soda turns the colour of the aqueous solution pale brown. This colouring matter is sent out nearly chemically pure.—*Dyes* wool and silk in acid baths a fine scarlet. Also known as New Coccin R (ACTIENGESELLSCHAFT) and Crystal Scarlet B (BADISCHE).

Crystal Violet—

(BADISCHE. SOCY. CHEM. IND., BASLE).— Hydrochlorate of hexamethyl-para-rosaniline

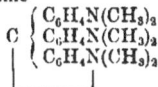

obtained in several ways. 1st. Acting with dimethylaniline upon tetramethyl-diamido-benzophenone chloride. 2nd. By direct action of phosgene in the presence of chloride of zinc or of thiophosgene upon dimethylaniline. 3rd. Condensation of tetramethyldiamido - benzhydrol with dimethylaniline and oxidation of the leuco-base which is produced.—*Dyes* wool and silk in neutral baths and cotton with a tannin mordant fine bright shades of violet, not fast to light or washing. (1883.) CARO AND KERN. *English Patents* 4,428, 1883; 5,450, 1883; 5,038, 1884; 11,030, 1884; 11,159, 1884.—*Bronzy or gold-green lustrous crystals,* soluble in water to a violet solution, in strong sulphuric acid to a yellow solution; on diluting with water the colour turns first green, then blue, lastly violet; soluble in alcohol. Hydrochloric acid changes the colour of the solution first blue, then green, finally yellow; caustic soda gives a violet precipitate.

Cumene—

Trimethyl benzene

$$C_9H_{12} = C_6H_3(CH_3)_3$$

is the fourth member of the benzene series of aromatic hydrocarbons, and is found present in coal tar; it boils at 166° C.; in other respects it resembles benzene. It is used only to make a few colouring matters. Like xylene, it occurs in three modifications, two only of which are known.

Cumidine—

Amido cumene

$$C_6H_2(CH_3)_3NH_2$$

Six isomeric compounds of this formula are possible. Commercial cumidine is obtained by heating the hydrochlorate of xylidine with methyl alcohol. It boils at about 225° C., and forms crystalline salts with acids. It is not much used in making colours.

Cumidine Red—

ACTIENGESELLSCHAFT.) —Syn. of Ponceau 3R.

Cumidine Scarlet—

(ACTIENGESELLSCHAFT.) – Syn. of Ponceau 3R.

Curcumein—

(ACTIENGESELLSCHAFT.)– A mixture of nitro-diphenylamine orange and nitro-diphenylamine obtained by treating diphenylamine orange with nitric acid. (1880.) KNECHT.— *Ochre-yellow powder,* slightly soluble in cold water, freely soluble in hot water; in strong sulphuric acid to a red-violet solution; on diluting with water this turns olive-brown. Hydrochloric acid added to the aqueous solution turns it crimson, caustic soda turns it yellow-brown.—*Dyes* wool in acid baths bright greenish yellow; also known as New Yellow, Citronin. This colouring matter is very similar to Indian Yellow (BAYER).

Curcumein—Syn. of Orange N.

Curcumine S—

LEONHARDT.) —

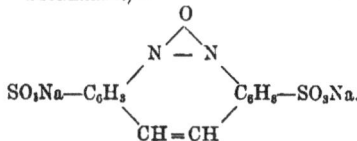

F

Sodium salt of azoxystilbene disulphonic acid, obtained by treating *p*-nitrotoluolsulphonic acid with caustic soda. (1883.)— *Brown powder,* dissolves in water to a brown-yellow solution, in strong sulphuric acid to a violet solution ; on diluting with water the solution turns yellow.—*Dyes* wool and silk in faintly acid baths yellows of reddish tones, which are fast. Known also as Sun Yellow, Maize.

Cyanamine—

$$N(CH_3)_2 . C_6H_4 . \overset{O}{N} . C_{10}H_5 . \overset{C_6H_4}{N} . N(CH_3)_2$$
$$OH$$

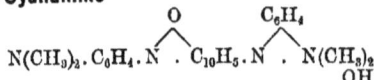

obtained by heating *B*-naphthol violet with alcoholic calcium hydroxide. (1890.) WITT. *Chem. Socy. Jour.* Abstracts, 1890, p. 1,307.—*Blackish brown plates,* soluble only in chloroform to a reddish violet solution. Soluble in acids to brownish orange solutions, which on diluting turn blue.

Cyanine—

(WILLIAMS, THOMAS AND DOWER.)—

$$C_{29}H_{35}N_2I$$

Quinoline and lepidin are mixed in equivalent quantities, acted on by amyl iodide and then by caustic alkali. (1860.) —GREVILLE WILLIAMS.— *Green lustrous crystals,* insoluble in cold, soluble in warm water with a pale blue colour, the solution has an odour of quinoline; in strong sulphuric acid to a colourless solution, which on heating evolves iodine; on diluting the acid solution with water it still remains colourless. Acids discharge the colour of the aqueous solution ; caustic soda gives a blue precipitate, changing to brown on warming. Exceedingly sensitive to light; a few hours' exposure will completely decolourise it. Not now used in dyeing; finds a use in isochromatic photography. Also known as Quinoline Blue.

Cyanosin—

Alkali salt of tetrabrom-dichlor-fluorescein-methylether

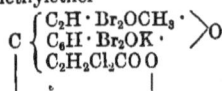

$$C \begin{cases} C_2H \cdot Br_2OCH_3 \cdot \\ C_6H \cdot Br_2OK \cdot \\ C_2H_2Cl_2CO \, O \end{cases} \!\!\! \Big\rangle O$$

obtained by methylating Phloxin P. (1876).—*Brown-red powder,* insoluble in cold, slightly soluble in boiling water, soluble in alcohol to a bluish red solution, which has an orange fluorescence, soluble in strong sulphuric acid to a yellow solution ; on warming bromine is evolved; on diluting with water a red-brown precipi-

tate is obtained. Hydrochloric acid added to the alcoholic solution destroys the fluorescence ; caustic soda has little action.

Cyanosin B—

(SOCY. CHEM. IND., BASLE.)—Sodium salt of tetrabrom-tetrachlor-fluorescein-ethylether

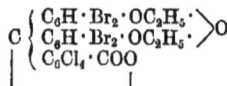

$$C \begin{cases} C_6H \cdot Br_2 \cdot OC_2H_5 \cdot \\ C_6H \cdot Br_2 \cdot OC_2H_5 \cdot \\ C_6Cl_4 \cdot COO \end{cases} \!\!\! \Big\rangle O$$

made by ethylating tetrabrom-tetrachlor fluorescein. (1882.) — *Red crystalline powder,* slightly soluble in water to a scarlet solution having yellow fluorescence, in strong sulphuric acid to a brown solution ; on diluting with water a brown-red flocculent precipitate is obtained. Hydrochloric acid gives a brown-red precipitate in the aqueous solution ; caustic soda turns the solution darker.—*Dyes* wool and silk very bright bluish scarlets from weak acid baths.

Cyclamine—

(GILLIARD, P. MONNET, AND CARTIER.)— Thio-tetra-iododichlorofluoresceine, obtained by treating dichlorofluoresceine with sodium sulphide and iodising the product. (1890).—*Bluish pink powder,* soluble in water to a bluish pink solution, also soluble in alcohol ; acetic acid causes the formation of a white precipitate, soluble with effervescence in strong sulphuric acid to an orange-brown solution ; on dilution with water a bluish pink precipitate falls down. Addition of hydrochloric acid to the aqueous solution throws down a deep pink precipitate; caustic soda has no action.—*Dyes* wool and silk from acid baths a fine bluish pink, resembling phloxine. Acids turn the colour a pale orange-scarlet ; caustic soda has no action. Bleeds strongly when boiled with soap.

D

Dahlia—

`Syn. of Hofmanns Violet.

Deep Red—

(CASTELHAZ.)—Syn. of Orcellin (HENRIET ROMAN AND VIGNON).

Deltapurpurin 5B—

(BAYER. ACTIENGESELLSCHAFT.)—

$$\text{Tetrazoditolyl} \!\! < \!\! \begin{array}{l} \textit{B}\text{-naphthylamine mono-} \\ \text{sulphonate } (\textit{delta}) \\ \textit{B}\text{-naphthylamine mono-} \\ \text{sulphonate } Br. \end{array}$$

(1886.) *English Patent* 5,846, 1886.— *Dark brown-red powder*, slightly soluble in water to a scarlet solution, in strong sulphuric acid to a blue solution; on diluting with water a reddish brown precipitate is obtained. Hydrochloric acid added to the aqueous solution gives a brown precipitate; caustic soda produces no change.—*Dyes* cotton from a soap bath fine scarlets; slightly brighter and bluer than Benzopurpurin 4B; not fast to light, turned brownish by acids; alkalies and soap have no action. Dyes wool from salt bath rather duller shades than on cotton. Dyes silk from a neutral soap bath pale pinkish shades. This colour is isomeric with Diamine Red 8B.

Deltapurpurin 7B—

(BAYER. ACTIENGESELLSCHAFT.)—

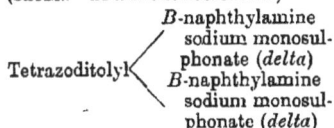

Tetrazoditolyl
{
  B-naphthylamine sodium monosulphonate (*delta*)
  B-naphthylamine sodium monosulphonate (*delta*)
}

(1886.)—*Dark brownish red powder*, very slightly soluble in cold water, easily soluble in hot water, in strong sulphuric acid to a blue solution; on diluting with water a brown precipitate is obtained. Acids added to the aqueous solution give a brown precipitate. Caustic soda has no action. —*Dyes* cotton from a soap bath deep shades of bluish scarlet, dulled a little by acids, faster to light than the benzopurpurins. Dyes wool from a salt bath deep shades of scarlet. Dyes silk from a neutral soap bath.

Deltapurpurin G—

(BAYER.)—

Tetrazodiphenyl
{
  B-naphthylamine sodium *delta*-monosulphonate
  B-naphthylamine sodium *delta*-monosulphonate
}

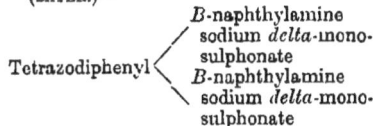

(1886.) *English Patent* 5,846, 1886.— *Brickish red powder*, very little soluble in cold water, soluble in hot water to a scarlet solution, in strong sulphuric acid to a blue solution; on diluting with water a dark brown precipitate is obtained. Hydrochloric acid added to the aqueous solution also gives a dark brown precipitate; acetic acid gives a brownish red precipitate; caustic soda added to the aqueous solution has no action.—*Dyes* cotton in a soap bath a bright scarlet of a yellow tone, not fast to acids and light. Dyes cotton in a salt bath brownish red shades. This colour is not now in use.

Diamine Black BO—

(CASSELLA AND CO.)—

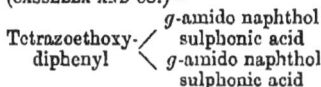

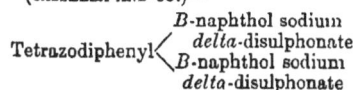

Tetrazoethoxy-diphenyl
{
  g-amido naphthol sulphonic acid
  g-amido naphthol sulphonic acid
}

(1891.) *English Patent* 16,699, 1889.— *Greyish black powder*, soluble in water to a dark blue solution, in alcohol to a purple solution, in acetic acid to a blue solution, in strong sulphuric acid to a blue solution; on diluting with water it turns purple. Hydrochloric acid added to the aqueous solution gives a reddish blue precipitate; caustic soda has no action.—*Dyes* cotton from a neutral soap bath deep navy blue shades, turned darker by acids; caustic soda turns them blue-black. Fast to boiling in soap and water. Diamine Black BO and RO (see below) when dyed on cotton, passed through a diazotising bath of sodium nitrite and hydrochloric acid, then into baths of naphthol, naphthylamine-ether, resorcin, etc., give new and fast shades of blue and black; the first and second give blues, the last gives blacks.

Diamine Black RO—

(CASSELLA.)—

Tetrazodiphenyl
{
  g-amido-naphthol sodium sulphonate
  g-amido-naphthol sodium sulphonate
}

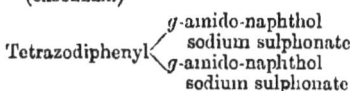

obtained by combining tetrazodiphenyl chloride with *gamma*-amido-naphthol sulphonic acid (1890.) *English Patent* 16,699, 1899.—*Grey-black powder*, soluble in water to a dark blue solution, in alcohol to a pale wine-red solution, in strong sulphuric acid to a bright blue solution; on diluting with water a bright blue precipitate is obtained. Hydrochloric acid added to the aqueous solution gives a bright blue precipitate; caustic soda a purple precipitate. —*Dyes* cotton blue to bluish black shades from a neutral soap bath, fast to acids and alkalies, moderately so to light and soaping.

Diamine Blue B—

(CASSELLA AND CO.)—

Tetrazodiphenyl
{
  B-naphthol sodium *delta*-disulphonate
  B-naphthol sodium *delta*-disulphonate
}

The naphthol acidjis obtained from Cassella's F acid. (1888.) *English Patent* 14,464, 1887.—*Dark brown powder*, soluble in water to a blue solution, in alcohol to a violet solution; insoluble in acetic acid, soluble in strong sulphuric acid to a pink solution; on adding water the colour turns yellow. Hydrochloric acid added to the aqueous solution has no action; caustic soda turns it violet.—*Dyes* cotton from a neutral soap bath dark reddish shades of blue, fast to acids, reddened by caustic soda, moderately fast to soaping and light.

Diamine Blue 6G—

(CASSELLA AND CO.) — Naphthylamine disulphonic acid - azo - ethoxynaphthyl - amine-azo-*delta*-naphthol sodium disulphonate

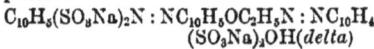

$$C_{10}H_5(SO_3Na)_2N : NC_{10}H_5OC_2H_5N : NC_{10}H_4 (SO_3Na)_4OH(delta)$$

obtained by diazotising naphthylamine disulphonic acid, combining with ethoxynaphthylamine, diazotising the body thus produced, and combining with naphthol *delta* disulphonic acid. (1891.) *English Patent* 7,087, 1889.—*Dark blackish violet powder*, with a bronzy lustre, soluble in water and alcohol to a blue solution, in acetic acid to a dark greenish blue solution, in strong sulphuric acid to a blackish brown solution, which on diluting with water turns a fawn-brown colour. Hydrochloric acid added to the aqueous solution gives a blue precipitate; caustic soda turns the colour paler, and gives a colourless precipitate.—*Dyes* cotton or wool from a neutral soap bath bright blue, turned dark greenish blue by strong hydrochloric acid, reddish blue by caustic soda. Not fast to boiling in strong soap and water.

Diamine Blue 3R—

(CASSELLA.) — (1889.) *English Patent* 14,464, 1887.—*Blackish blue powder*, soluble in water to a reddish blue solution, in strong sulphuric acid to a blue solution; on diluting with water a purple precipitate is obtained. Hydrochloric acid gives a purplish precipitate. Caustic soda turns the colour of the aqueous solution scarlet.—*Dyes* cotton from a neutral soap bath reddish blue, fast to light, acids, and washing; alkalies redden the shades.

Diamine Blue Black E—

(CASSELLA AND CO.)—

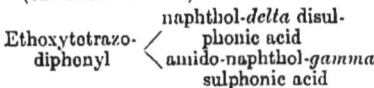

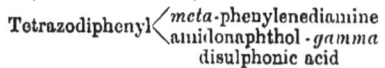

naphthol-*delta* disulEthoxytetrazo- / phonic acid
diphenyl \ amido-naphthol-*gamma* sulphonic acid

(1891.) *English Patent* 16,699, 1889.— *Greyish black powder*, soluble in water to a deep blue solution, in alcohol to a purple solution, in acetic acid to a violet solution, in strong sulphuric acid to a bright blue solution, from which on diluting with water a red-violet precipitate falls out. Hydrochloric acid added to the aqueous solution turns it a bright blue.—*Dyes* cotton or wool from neutral soap baths navy

blues, fast to acids and soaping; caustic soda turns the shade darker. Cotton dyed with Diamine Blue Black E can be diazotised with bath of sodium nitrite, and dyed new shades of blue by passing into developing baths containing naphthol, naphthylamine, ether, or resorcin.

Diamine Brown V—

(CASSELLA AND CO.)—

Tetrazodiphenyl⟨*meta*-phenylenediamine / amidonaphthol -*gamma* disulphonic acid

(1891.) *English Patent* 16,699, 1889.— *Blackish brown powder*, insoluble in cold water, soluble in hot water to a wine-red solution, in alcohol to a deep red solution, insoluble in acetic acid, soluble in strong sulphuric acid to a blue solution, which gives a brown-violet precipitate on adding water. Hydrochloric acid added to the aqueous solution gives a brown-violet precipitate; caustic soda has no action.— *Dyes* cotton from a neutral soap bath deep shades of brown-violet, turned a little bluer by acids, brown by caustic soda, fast to soaping. Cotton dyed with Diamine Brown V can be re-dyed by first passing the cotton into a diazotising bath, then into a developing bath of naphthol, resorcin, naphthylamine or chrysoidine, giving deep browns, fast to soaping, acids and alkalies.

Diamine Fast Red F—

(CASSELLA.)—

Tetrazodiphenyl⟨sodium salicylate / *p*-amido-naphthol sodium sulphonate

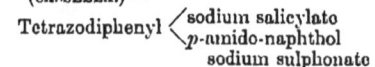

(1890.)—*Brownish red powder*, soluble in water to a red solution, in strong sulphuric acid to a bright blue solution; on diluting with water a dark crimson precipitate is obtained. Hydrochloric acid added to the aqueous solution gives a dark bluish red precipitate; caustic soda gives a red precipitate.—*Dyes* cotton from a soap bath dull bluish red; wool from a salt bath bright crimson; acids turn it blackish blue, alkalies darken it; moderately fast to soaping. Faster to light on wool than on cotton.

Diamine Red B—

(ACTIENGESELLSCHAFT.) — 1890. — *Red powder*, soluble in water to an orange-red solution, soluble in alcohol to a reddish orange, in acetic acid to a brownish red, in strong sulphuric acid to an olive-yellow solution, which on diluting with water changes to a lemon-yellow. Hydrochloric acid added to the aqueous solution gives a brown-violet precipitate; caustic soda has no action.—*Dyes* cotton from an alkaline soap bath deep shades of blue, fast to acids and washing; alkalies redden it, not fast to light.

Diamine Red 3B—

(CASSELLA. ACTIENGESELLSCHAFT.)—

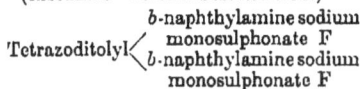

$$\text{Tetrazoditolyl} \Big\langle \begin{array}{l} b\text{-naphthylamine sodium} \\ \text{monosulphonate F} \\ b\text{-naphthylamine sodium} \\ \text{monosulphonate F} \end{array}$$

(1886.) *English Patent* 12,908, 1886.— *Dark brown-red powder*, slightly soluble in water to a red solution, in strong sulphuric acid to a blue solution; on diluting with water a brownish red precipitate is obtained. Hydrochloric acid added to the aqueous solution gives a dark brown precipitate; caustic soda gives a red precipitate.—*Dyes* cotton from a soap bath bluish scarlet shades, resembling those from Benzopurpurine 6B. Acids turn them a purplish red, alkalies have no action. Dyes wool fine scarlet from a salt bath. Dyes silk fine scarlet from a neutral soap bath. This colour is said to be isomeric with Deltapurpurine 5B, but it gives rather bluer shades. It is not now sent out by Messrs. Cassella and Co.

Diamine Red N—

(CASSELLA.)—(1889.) *English Patent* 14,464, 1887.—*Dark red powder*, slightly soluble in water, soluble in strong sulphuric acid to a blue solution; on diluting with water a dark blue precipitate is obtained. Acids throw down a dark purplish precipitate from the aqueous solution. Caustic soda brightens the solution.—*Dyes* cotton from an alkaline soap bath bluish red, not fast to light; acids turn it blue; fast to washing.

Diamine Scarlet B—

(SOCIÉTÉ LYONNAISE.)—

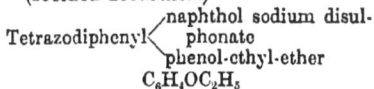

$$\text{Tetrazodiphenyl} \Big\langle \begin{array}{l} \text{naphthol sodium disul-} \\ \text{phonate} \\ \text{phenol-ethyl-ether} \\ C_6H_4OC_2H_5 \end{array}$$

(1890.) *English Patent* 12,560, 1889.— *Scarlet powder*, slightly soluble in cold water, easily in hot water to a scarlet solution, in alcohol to a pale pink, in acetic acid to a scarlet solution, in strong sulphuric acid to a violet solution, which on diluting with water turns a brownish orange. Hydrochloric acid added to the aqueous solution gives a brown-red precipitate; caustic soda darkens the colour. —*Dyes* cotton, wool and silk from neutral soap baths fine fiery scarlets, turned brown by acids and alkalies, not fast to strong soaping, turns yellower on exposure to light.

Diamine Violet N—

(CASSELLA.)—

$$\text{Tetrahydrazin-} \Big\langle \begin{array}{l} \text{amido-naphthol} \\ \text{amido-naphthol} \end{array}$$
diphenyl

$$C_6H_4 \cdot NH \cdot N \cdot C_{10}H_5NH$$
$$\qquad\qquad\qquad OH$$
$$C_6H_4 \cdot NH \cdot N \cdot C_{10}H_5NH$$
$$\qquad\qquad\qquad OH$$

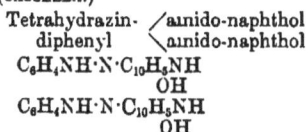

(1890.)—*Purplish brown powder*, soluble in water to a purple solution, in strong sulphuric acid to a blue solution; on diluting with water a red precipitate is obtained. Hydrochloric acid gives a blue precipitate in the aqueous solution; caustic soda turns it a little bluer.—*Dyes* cotton from a soap bath a puce colour; wool from a salt bath a violet-red; turned blue by acids, fast to alkalies and soaping, moderately fast to light.

Diamine Yellow N—

(CASSELLA.)—

$$\text{Tetrazoethoxy-} \Big\langle \begin{array}{l} \text{sodium salicylate} \\ \text{phenetol} \end{array}$$
diphenyl

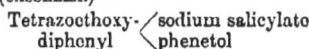

(1889.) *English Patent* 14,464, 1887.— *Yellowish brown paste*, very slightly soluble in cold, easily in hot water; soluble in strong sulphuric acid to a magenta solution; on diluting with water a brown precipitate is obtained. Hydrochloric acid gives a brown precipitate in the aqueous solution; caustic soda turns it redder.— *Dyes* cotton from a neutral soap bath orange-yellows, fast to light, washing, reddened slightly by soaping, turned brown by acids.

Diamond Black—

(BAYER.)—

$$C_6H_5 \begin{cases} OH \\ N:NC_{10}H_6N:NC_{10}H_5OHSO_3Na \\ COOH \end{cases}$$

(1890.)—*Brown-black powder*, soluble in water to a purple solution, in strong sulphuric acid to a green solution; on diluting with water a brown precipitate is obtained. Hydrochloric acid added to the aqueous solution gives a claret-red precipitate; caustic soda turns it blue.—*Dyes* wool from an acid bath bluish black shades, fast to acids, alkalies, soaping and light.

Diamond Green—

Syn. of Malachite Green.

Diamond Green—

(BAYER.)—Salicylic acid - azo - naphtha-lene-azo-dioxynaphthalene sodium mono-sulphonate

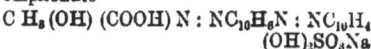

$$C H_3(OH)(COOH) N : NC_{10}H_6N : NC_{10}H_4$$
$$\qquad\qquad\qquad\qquad (OH)_2SO_3Na$$

(1890.)—*Bronzy black powder*, slightly soluble in cold, readily soluble in hot water to a dark navy-blue solution, in alcohol to a violet-blue, slightly soluble in acetic acid to a violet-blue solution, in strong sulphuric acid to a green solution, from which on diluting with water a blackish blue precipitate falls out. Hydrochloric acid added to the aqueous solution throws down a blackish blue precipitate; caustic soda turns the colour of the solution a violet-blue. — *Dyes* chrome mordanted wool and silk deep olive shades of green, fast to acids, turned brown by caustic soda, fast to soaping and light.

Diamond Yellow G—

(BAYER.)—Metamido-benzoic acid-azo-salicylic acid

$$\left\{ \begin{array}{l} C_6H_4 \\ COONa \end{array} \right. N:N \left\{ \begin{array}{l} C_6H_3OH \\ COONa \end{array} \right.$$

(1891.)—*Yellowish brown paste*, insoluble in cold water, soluble in hot water and alcohol to a dark amber solution, insoluble in acetic acid. Soluble in strong sulphuric acid to a Bismarck brown solution, from which on diluting with water the original colouring matter is thrown down. Soluble in caustic soda to a very deep amber-coloured solution.—*Dyes* chrome mordanted wool olive shades of yellow, turned scarlet-red by strong hydrochloric acid, reddish by caustic soda, fast to boiling in soap and water, fast to light.

Diamond Yellow R—

(BAYER.)—Orthoamido-benzoic acid-azo-salicylic acid

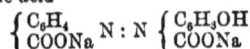

$$\left\{ \begin{array}{l} C_6H_4 \\ COONa \end{array} \right. N:N \left\{ \begin{array}{l} C_6H_3OH \\ COONa \end{array} \right.$$

(1891).—Has similar properties to Diamond Yellow G, but dyes rather redder shades of yellow.

Dianthine B—

Syn. of Erythrosine.

Dianthine G—

Syn. of Erythrosine G.

Dioxine—

(LEONHARDT.)—Nitrosodioxynaphthaline

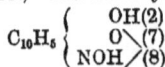

$$C_{10}H_5 \left\{ \begin{array}{l} OH(2) \\ O\backslash(7) \\ NOH \diagup (8) \end{array} \right.$$

(1890.) *English Patent* 17,223, 1889.— *Dark red paste*, slightly soluble in boiling water to an olive-yellow solution, in strong sulphuric acid to a green solution. Hydrochloric acid added to the aqueous solution has no action; caustic soda turns it crimson.—*Dyes* wool, mordanted with chrome,

reddish brown; mordanted with iron green, with a mixed chrome, iron yellow brown. Fast to acids, alkalies, soaping and light.

Diphenyl Orange—

Syn. of Acid Yellow (1).

Diphenylamine Orange—

Syn. of Acid Yellow (1).

Diphenylamine Blue Spirit Soluble—

Hydrochlorate of triphenylpararosaniline (?), obtained by action of oxalic acid upon diphenylamine or by acting with carbon-hexachloride upon a mixture of Diphenylamine and Ditolylamine. (1866.) —*Brown powder*, insoluble in water, slightly soluble in cold, easily in warm alcohol, and in strong sulphuric acid to a brown solution. Used only for making the Alkali Blue D.

Double Brilliant Scarlet G—

(ACTIENGESELLSCHAFT.) — Sodium salt of *B*-naphthylamine sulphonic acid-azo-*B*-naphthol

$$C_{10}H_6SO_3NaN : NC_{10}H_6OH$$

obtained from *B*-naphthylamino monosulphonic acid Br., by diazotising and combining with *B*-naphthol. (1882.)— *Red-brown powder*, soluble in water to a scarlet solution, in strong sulphuric acid to a magenta-red solution; on diluting with water a brown-red precipitate is thrown down; addition of hydrochloric acid to the aqueous solution throws down a brown precipitate; caustic soda throws down a red-brown precipitate, soluble on adding much water.—*Dyes* wool in acid baths fine scarlet shades, fiery and solid, fairly fast. This colour is isomeric with Fast Brown 3B, Acid Ponceau, Fast Brown (BADISCHE), and Fast Red A.

Double Scarlet—

(KALLE.)—Syn. of Fast Scarlet (BADISCHE).

Double Scarlet S Extra—

(ACTIENGESELLSCHAFT.)—Sodium salt of *B*-naphthylamine monosulphonic acid-azo-*a*-naphthol monosulphonic acid

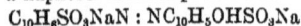

$$C_{10}H_6SO_3NaN : NC_{10}H_5OHSO_3Na$$

obtained by diazotising *B*-naphthylamine monosulphonic acid Br., and combining with *a*-naphthol monosulphonic acid NW. (1882.)—*Brown-red powder*, soluble in water to a scarlet solution, in strong sulphuric acid to a crimson solution; on diluting with water this turns scarlet. Addition of hydrochloric acid to the aqueous solution gives a yellow-brown flocculent precipitate. Caustic soda does not change the colour of the aqueous solution.—*Dyes* wool in acid baths fine scarlets, fast. Also known as Brilliant Ponceau (BAYER). This colour is isomeric with Pyrotin (DAHL).

E

Emeraldine—

A green colour produced directly on the fibre which has been impregnated with chlorate of potash by padding in a solution of aniline. CALVERT, LOWE AND CLIFT. *English Patent* 1,426, 1860.

English Yellow—

Syn. of Victoria Yellow.

Eosin—

(BADISCHE.)—Alkali salt of Tetrabromfluorescein

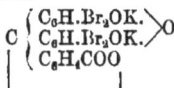

$$C \begin{cases} C_6H.Br_2OK. \\ C_6H.Br_2OK. \\ C_6H_4COO \end{cases} O$$

obtained by the bromination of fluorescein in alcoholic or aqueous solution. (1874.) CARO.—*Red crystalline powder with a bluish tone*, easily soluble in water to a bluish red solution; dilute solutions have a strong green fluorescence, soluble in alcohol to a blue-red solution having a yellow-green fluorescence. Soluble in strong sulphuric acid to a yellow solution; on diluting with water a scarlet precipitate is obtained. Hydrochloric acid added to the aqueous solution gives a flocculent scarlet precipitate ; caustic soda has little action. Eosin is precipitated by solutions of lead acetate, alumina sulphate. —*Dyes* wool and silk from acetic acid baths fine rose to scarlet shades, having a fluorescent appearance. Fast to washing, but not to light. Known also as Eosin A, Eosin Yellowish, Eosin GGF, Water Soluble Eosin, Eosin B, Eosin JL.

Eosin A—

(BADISCHE.)—Syn. of Eosin.

Eosin B—

Syn. of Eosin.

Eosin BB—

(SOCY. CHEM. IND., BASLE.)—Is the potassium salt of tetrabromfluorescein-ethylether, and is the homologue of, and has similar properties to, Erythrin.

Eosin BN—

(BADISCHE.)—Alkali salt of bromdinitrofluorescein

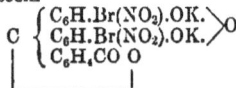

$$C \begin{cases} C_6H.Br(NO_2).OK. \\ C_6H.Br(NO_2).OK. \\ C_6H_4CO \ O \end{cases} O$$

obtained by nitrating dibromofluorescein in an aqueous solution, or by brominating dinitrofluorescein in an alcoholic solution. (1875.)—CARO. *Brown-red crystalline powder*, easily soluble in water to a scarlet solution, having a weak green fluorescence; soluble in strong sulphuric acid to a yellow-brown solution, on heating which bromine is evolved; on diluting with water a brown-yellow precipitate is obtained. Adding hydrochloric acid to the aqueous solution gives a yellow-green flocculent precipitate. Caustic soda has no action. Precipitated by solutions of lead acetate, alumina sulphate.—*Dyes* silk and wool from weak acid baths bright bluish scarlet shades those on silk are weakly fluorescent. Fast to acids and washing, but not to light. Known also as Methyleosin, Safrosin, Eosin BW, Scarlet J, 2J.V, Eosin Scarlet B.

Eosin Blue Shade—

Syn. of Erythrosine.

Eosin BW—

Syn. of Eosin BN.

Eosin GGF—

(CASSELLA.)—Syn. of Eosin.

Eosin 3G—

(CASSELLA.)—A mixture of dibrom- and tetra-bromfluorescein.

Eosin J—

(BADISCHE.)—Syn. of Erythrosine.

Eosin Orange—

(SOCY. CHEM. IND., BASLE.)—A mixture of dibrom- and tetra-bromfluorescein.

Eosin S—

(BADISCHE.)—Syn. of Primrose.

Eosin Scarlet B—

(CASSELLA.)—Syn. of Eosin BN.

Eosin Spirit Soluble—

(BADISCHE.)—Syn. of Erythrine.

Eosin Spirit—

Syn. of Primrose.

Eosin Water Soluble—

(MEISTER, LUCIUS.)—Syn. of Eosin.

Eosin Yellow Shade—

(ACTIENGESELLSCHAFT.)—Syn. of Eosin.

Erika B—

(ACTIENGESELLSCHAFT.)—A derivative of thioxylidine. (1889.) *English Patent* 17,333, 1888.—*Dark brownish red powder,* slightly soluble in cold, easily soluble in boiling water to a crimson solution, soluble in alcohol to a pale crimson, in acetic acid to a red solution, in strong sulphuric acid to a crimson solution; on diluting with water an orange-yellow precipitate falls out. Hydrochloric acid added to the aqueous solution gives a red flocculent precipitate, while caustic soda gives a purple precipitate.—*Dyes* cotton, wool and silk from a salt bath, with small quantities of dyestuff, safranine shades of pink, fast to soaping and dilute acids; strong acid reddens the shade, alkalies darken it.

Erika G—

(ACTIENGESELLSCHAFT.)—A derivative of thioxylidine resembling Titan pink in its composition. (1889.) *English Patent* 17,333, 1888.—*Brownish red powder,* slightly soluble in cold, easily soluble in hot water to a red solution, in alcohol to a bluish scarlet, in acetic acid to a red solution, in strong sulphuric acid to a crimson solution, from which on diluting with water a yellow precipitate falls out. Hydrochloric acid added to the aqueous solution throws down a red flocculent precipitate, while caustic soda throws down a purple precipitate. —*Dyes* cotton, wool and silk from a salt bath, with small quantities of dyestuff, fine safranine pinks, fast to washing and dilute acids ; strong acids redden the shade, caustic soda darkens it, moderately fast to light.

Erythrin—

(BADISCHE.)—Potassium salt of tetra-bromfluorescein-methyl-ether

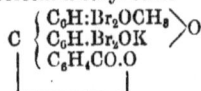

$$C \begin{cases} C_6H{:}Br_2OCH_3 \\ C_6H.Br_2OK \\ C_6H_4CO.O \end{cases} \!\!> O$$

obtained by methylating Eosin. (1874.) CARO. *Greenish lustrous powder or crystals,* slightly soluble in cold water, easily in hot water to a scarlet solution ; soluble in alcohol to a red solution, having a brownish yellow fluorescence; soluble in strong sulphuric acid to a yellow solution ; on heating this gives off bromine. On diluting with water a brownish yellow precipitate is obtained. Addition of hydrochloric acid to the aqueous solution gives a brownish yellow precipitate. Caustic soda turns the colour of the solution darker, and gives a dark green fluorescence. —*Dyes* wool and silk from acetic acid baths bluish scarlets, very brilliant ; on silk they have a dark reddish fluorescence. The shades are fast to acids and washing, but not to light. Known also as Spirit Soluble Eosin, Spirit Primrose, Methyl Eosin.

Erythrin X—

(BADISCHE.)—Syn. of Ponceau 5R.

Erythrosine—

(BENDSCHEDLER)—Alkali salt of tetraiod-fluorescein

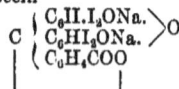

$$C \begin{cases} C_6H.I_2ONa. \\ C_6HI_2ONa. \\ C_6H_4COO \end{cases} \!\!> O$$

obtained by iodising fluorescein. (1876).— *Brown powder,* soluble in water to a scarlet solution which has little fluorescence; in strong sulphuric acid to a brown yellow solution, which on heating gives off iodine ; on diluting with water a brown-yellow precipitate is obtained. Hydrochloric acid added to the aqueous solution gives a brown-yellow precipitate ; caustic soda causes little change. Precipitated by solutions of lead acetate, alumina sulphate. —*Dyes* wool and silk bluish scarlet shades, very bright, fast to acids and washing, not to light. Known also as Erythrosine B, Pyrosine B, Iodine Eosine B, Rose B, Dianthine B, Soluble Primrose, Eosine J, Eosine BA.

Erythrosine G—

Alkali salt of diiodfluorescein

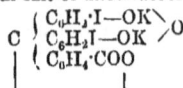

$$C \begin{cases} C_6H_2I \cdot I{-}OK \\ C_6H_2I{-}OK \\ C_6H_4 \cdot COO \end{cases} \!\!> O$$

obtained by treating fluorescein with iodine and hydriodic acid, or iodine chloride and alkali. (1875.) — *Yellow-brown powder,* soluble in water to a bright scarlet solution which has little fluorescence ; in strong sulphuric acid to a brown-yellow solution, which on heating evolves iodine ; on diluting with water a brown-yellow precipitate is obtained. Hydrochloric acid added to the aqueous solution gives a brown-yellow precipitate; caustic soda has no action. Precipitated by solutions of lead acetate, alumina sulphate.—*Dyes* wool and silk from weak acid baths bright scarlets, having a red fluorescence. Fast to acids and washing, not to light. Known also as Dianthine G, Pyrosine J, Iodine Eosine G.

Ethyl Eosin—

Syn. of Primrose.

Ethyl Green—

(READ HOLLIDAY AND SONS). — Zinc chloride double salt of Bromethyl-hexamethyl-para-rosaniline hydrochlorate

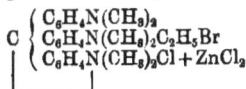

$$C \begin{cases} C_6H_4N(CH_3)_2 \\ C_6H_4N(CH_3)_2C_2H_5Br \\ C_6H_4N(CH_3)_2Cl + ZnCl_2 \end{cases}$$

obtained by acting with ethylbromide on methyl violet. (1866.)—HOLLIDAY. *English Patent*, 10 May, 1866.—*Bright green crystalline powder*, easily soluble in water to a greenish blue solution, in strong sulphuric acid to a yellow solution; on diluting with much water this turns green. Hydrochloric acid added to the aqueous solution turns it first green, then yellow; caustic soda discharges the colour of the solution.—*Dyes* wool and silk in neutral baths; cotton requires a mordant of tannin. Bluish green shades are obtained, only moderately fast to light and washing. Also known as Methyl Green (BAYER).

Ethyl Green—

Syn. of Brilliant Green.

Ethyl Violet—

(BADISCHE. SOCY. CHEM. IND., DASLE). —Hydrochlorate of hexaethyl - pararosaniline

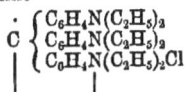

$$C \begin{cases} C_6H_4N(C_2H_5)_2 \\ C_6H_4N(C_2H_5)_2 \\ C_6H_4N(C_2H_5)_2Cl \end{cases}$$

obtained by acting with diethylaniline upon tetraethyldiamido - benzophenone chlorid. (1883.)—KERN AND CARO. *English Patents* 4,428, 1883; 5,450, 1883; 5,038, 1884; 11,030, 1884; 11,159, 1884.—*Green crystalline powder*, easily soluble in water to a bluish violet solution, in strong sulphuric acid to a brownish-yellow solution; on diluting with water turns green. Hydrochloric acid turns the aqueous solution reddish yellow; caustic soda gives a grey-violet precipitate.—*Dyes* wool and silk in neutral baths. Cotton requires a mordant of tannin and antimony. Bluish violet shades are given, which are not fast to light or washing.

Eurhodine—

A body obtained by heating orthoamidoazotoluene with naphthylamine hydrochloride, having the formula $C_{17}H_{19}N_3$, discovered by Witt in 1883. Has basic properties, forms three series of salts, the monoacid salts being the most stable, the others are decomposed by water. Although the normal salts have strong colouring power, dyeing fibres bright scarlets, they are decomposed by water giving the free Eurhodine base; and as this change takes place on the dyed fibres by simply washing (the colour turning from scarlet to yellow), the Eurhodine has not come into practical use as a dye. Neutral Red and Neutral Violet belong to the Eurhodine group of colouring matters.

F

Fast Blue R—

(ACTIENGESELLSCHAFT.)—Syn. of Induline Water Soluble.

Fast Blue R Spirit Soluble—

Syn. of Induline Spirit Soluble.

Fast Brown—

(BADISCHE.)—Sodium salt of naphthionic acid-azo-a-naphthol

$$C_{10}H_6SO_3Na \ N : NC_{10}H_6OH$$

obtained by diazotising naphthionic acid and combining with a-naphthol. (1878).— *Dark brown powder*, soluble in water to a yellow-brown solution, in strong sulphuric acid to a blue solution; on diluting with water a brown precipitate forms. Addition of hydrochloric acid to the aqueous solution throws down a brown precipitate; caustic soda turns the colour of the aqueous solution reddish brown.— *Dyes* wool in acid baths reddish brown, fairly fast to acids, washing and light. This colour is isomeric with Fast Red A.

Fast Brown –

(BAYER.)— Sodium salt of naphthionic acid disazo-resorcin

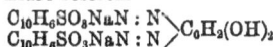

$$\begin{matrix} C_{10}H_6SO_3NaN : N \\ C_{10}H_6SO_3NaN : N \end{matrix} \Big\rangle C_6H_2(OH)_2$$

obtained by diazotising two molecules of naphthionic acid and combining with one molecule of resorcin (1881).

Fast Brown—

(MEISTER, LUCIUS AND BRUNING.)— Sodium salt of xylene sulphonic acid-disazo-a-naphthol

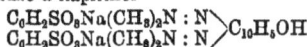

$$\begin{matrix} C_6H_2SO_3Na(CH_3)_2N : N \\ C_6H_2SO_3Na(CH_3)_2N : N \end{matrix} \Big\rangle C_{10}H_6OH$$

obtained by diazotising two molecules of xylidine monosulphonic acid and combining with one molecule of a-naphthol.— *Dark brown powder*, soluble in water to a brown solution, in strong sulphuric acid to a violet solution; on diluting with water this turns red. Addition of hydrochloric acid to the aqueous solution throws down a violet precipitate; caustic soda turns the aqueous solution a reddish

G

yellow.—*Dyes* wool and silk in acid baths a brownish red.

Fast Brown 3B—

(ACTIENGESELLSCHAFT.)—Sodium salt of *B*-naphthylamine sulphonic acid-azo-*a*-naphthol

$$C_{10}H_6SO_3NaN : NC_{10}H_6OH$$

obtained by diazotising *B*-naphthylamine monosulphonic acid Br. and combining with *a*-naphthol. (1882).—*Brown powder*, soluble in water to a brown solution, in strong sulphuric acid to a blue solution ; on diluting with water a red-violet precipitate forms. On adding hydrochloric acid to the aqueous solution it turns red-violet ; caustic soda turns the aqueous solution crimson.—*Dyes* wool in acid baths brown shades, moderately fast. This colouring matter is isomeric with Acid Ponceau, Double Brilliant Scarlet G, Fast Brown (BADISCHE), Fast Red A.

Fast Brown G—

(ACTIENGESELLSCHAFT.)—Sodium salt of disulphanilic acid disazo-*a*-naphthol

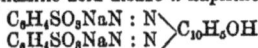

$$\left. \begin{array}{l} C_6H_4SO_3NaN : N \\ C_6H_4SO_3NaN : N \end{array} \right\rangle C_{10}H_5OH$$

obtained by diazotising sulphanilic acid and combining with *a*-naphthol (1882.)—*Brown powder*, soluble in water to a red-brown solution; in strong sulphuric acid to a violet solution; on diluting with water a yellow-brown solution is obtained. On adding hydrochloric acid to the aqueous solution a violet precipitate is obtained ; this is soluble in water to a brown solution or in dilute acid to a violet solution; caustic soda turns the colour of the aqueous solution a cherry-red.—*Dyes* wool in acid baths a reddish brown.

Fast Cotton Blue—

(ACTIENGESELLSCHAFT.)—Syn. of Meldolas Blue.

Fast Green—

Syn. of Malachite Green.

Fast Green—

(BAYER.)—Sodium salt of tetramethyl-di-benzyl - pseudorosaniline - disulphonic acid

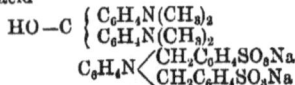

$$HO-C \left\{ \begin{array}{l} C_6H_4N(CH_3)_2 \\ C_6H_4N(CH_3)_2 \end{array} \right.$$
$$C_6H_4N \left\langle \begin{array}{l} CH_2C_6H_4SO_3Na \\ CH_2C_6H_4SO_3Na \end{array} \right.$$

obtained by acting with meta-nitrobenzaldehyde upon dimethylaniline, reduction of the condensation product thus obtained benzylating and sulphonising. (1885.) — *Dark blue - green crystalline powder*, soluble in water to a greenish

blue solution, in strong sulphuric acid to a yellowish red solution; on diluting with water it turns at first colourless, but on greater dilution turns greenish blue. Hydrochloric acid gives a yellow solution, turning on addition of water greenish yellow ; caustic soda on warming discharges the colour.—*Dyes* wool in acid baths, silk in a broken soap bath, dark greens, moderately fast to light, fast to washing.

Fast Green Blue—

(BAYER.)—(1891.) Sent out in the form of a violet-blue powder and of a liquid of two strengths, 10 per cent. and 20 per cent. The powder is soluble in water to a bright greenish blue solution, in alcohol and acetic acid to a greenish blue solution, in strong sulphuric acid to an amber-coloured solution, which on diluting with water turns yellow. Hydrochloric acid added to the aqueous solution turns it yellow ; caustic soda has no action.—*Dyes* wool from acid baths very bright bluish green shades, turned yellow by acids, fast to alkalies, not fast to soaping or light.

Fast Green Extra Blue—

(BAYER.) — A stronger brand of the above.

Fast Myrtle—

Syn. of Resorcin Green.

Fast Ponceau B—

(BADISCHE.)—Syn. of Biebrich Scarlet.

Fast Ponceau 2B—

(BADISCHE.)—Syn. of Ponceau S Extra.

Fast Red—

(ACTIENGESELLSCHAFT.)—Syn. of Fast Red E.

Fast Red A—

(BADISCHE.)—Sodium salt of naphthionic acid azo-*B*-naphthol

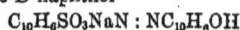

$$C_{10}H_6SO_3NaN : NC_{10}H_6OH$$

obtained by diazotising naphthionic acid and combining with *B*-naphthol. (1877.)—*Brown-red powder*, slightly soluble in cold water, more freely in hot water to a bright red solution, in strong sulphuric acid to a violet solution ; on diluting with water a yellow precipitate forms. Hydrochloric acid added to the aqueous solution throws down a yellow-brown precipitate ; caustic soda turns the aqueous solution darker.—*Dyes* wool and silk in acid baths dark reds, very fast to light, washing and acids. Also known as Roccellin, Orcellin No. 4, Rubidin.

Fast Red B—

(BADISCHE.)—Sodium salt of a-naphthyl-amine-azo-B-naphthol disulphonic acid

$$C_{10}H_7N:NC_{10}H_4OH(SO_3Na)_2$$

obtained by treating diazo-a-naphthyl-amine with B-naphthol disulphonic acid R. (1878.)—*Brown powder*, soluble in water with a magenta-red colour, in strong sulphuric acid to a blue solution, which on diluting with water turns a magenta-red; adding hydrochloric acid to the aqueous solution does not alter the colour, while caustic soda turns it a yellow-brown.— *Dyes* wool and silk in acid baths a bluish red, which is very fast to acids, etc. Also known as Bordeaux B (ACTIENGESELL-SCHAFT), Red C (BADISCHE).

Fast Red C—

Syn of Azorubin S.

Fast Red D—

(BADISCHE.)—Syn. of Amaranth.

Fast Red E—

(BADISCHE. BAYER.) — Sodium salt of naphthionic acid-azo-B-naphthol mono-sulphonic acid

$$C_{10}H_6SO_3NaN : NC_{10}H_5OHSO_3Na$$

obtained by diazotising naphthionic acid, and combining with B-naphthol mono-sulphonic acid S. (1878.)—*Red-brown powder*, soluble in water to a dark crimson-red solution, in strong sulphuric acid to a violet solution; on diluting with water a red solution is obtained. Hydrochloric acid does not change the colour of the aqueous solution, while caustic soda turns it brown.—*Dyes* wool in acid baths dark bluish reds. Also known as Fast Red (ACTIENGESELLSCHAFT), Caroubier. Iso-meric with Crocein 3BX and Azo Rubin S.

Fast Scarlet—

(BADISCHE.)—Sodium salt of benzene sul-phonic acid azo-benzene-azo-B-naphthol

$$C_6H_4SO_3NaN : NC_6H_4N : NC_{10}H_6OH$$

obtained by diazotising amido-azo-benzene monosulphonic acid, and combining with B-naphthol. (1878.) *English Patent* 5,004, 1879. — *Red-brown crystalline powder*, soluble in water with a scarlet colour, in strong sulphuric acid with a green colour ; on diluting with water this turns first blue, then blue-red, finally scarlet. Addition of hydrochloric acid to the aqueous solution, if weak, turns it yellower; if strong, gives a red flocculent precipitate. Caustic soda gives a brown flocculent precipitate.—*Dyes* wool in acid baths bright scarlets. Also known as Double Scarlet (KALLE).

Fast Violet Blue Shade--

(BAYER.) — Paratoluene sodium mono-sulphonate-azo-naphthalene-azo-naphthol sodium sulphonate

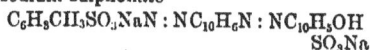

$$C_6H_5CH_3SO_3NaN : NC_{10}H_6N : NC_{10}H_5OH \\ SO_3Na$$

obtained by diazotising paratoluene sodium sulphonate-azo-naphthylamine and com-bining with naphthol sodium sulphonate. (1888.) — *Brown-violet powder* with a slight bronzy lustre, soluble in water to a violet solution, in alcohol to a deep crim-son solution, in acetic acid to a fine reddish violet solution, in strong sulphuric acid to an olive-green solution, which on diluting with water turns red-violet. Hydrochloric acid added to the aqueous solution throws down a violet precipitate; caustic soda throws down a violet precipitate.—*Dyes* wool in acid baths or on chrome mordants full deep red-violet, fast to light, air, washing and dilute acids.

Fast Violet Reddish—

(BAYER.)—Benzene-para-sodium mono-sulphonate azo-naphthalene-azo-naphthol sodium monosulphonate

$$C_6H_4SO_3NaN:NC_{10}H_6N:NC_{10}H_5OHSO_3Na$$

Dark green-black powder, slightly soluble in cold, easily in hot water to a magenta-coloured solution ; on first adding the water a brownish colour is developed; this changes on heating to the red. Very slightly soluble in alcohol, easily soluble in acetic acid to a magenta-coloured solu-tion, in strong sulphuric acid to a dark olive-green solution; on diluting with water a dark purplish red precipitate is obtained. Hydrochloric acid added to the aqueous solution throws down a dark crimson-red precipitate. Caustic soda turns the colour of the solution bluer.— *Dyes* wool from acid baths a dark red-violet, fast to acids and alkalies, to soap-ing and light.

Fast Yellow—

(BADISCHE.)—Sodium salt of amido-azo-benzene sulphonic acid. (1878.) — See Acid Yellow (2).

Fast Yellow Extra—

(BAYER.) — Sodium salt of amido-azo-benzene disulphonate. See Acid Yellow (2).

Fast Yellow G—

(KALLE.)—Syn. of Acid Yellow (2).

Flavaniline—

(MEISTER, LUCIUS).—Hydrochlorate of a-paramidophenyl-gamma-lepidine

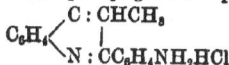

$$C_6H_4\begin{cases} C : CHCH_3 \\ N : CC_6H_4NH_2HCl \end{cases}$$

obtained by heating acetanilide with chloride of zinc to 270° C. (1881.)— *Orange-red crystalline powder*, soluble in water to a yellow solution, in strong sulphuric acid to a colourless or pale yellow solution with a blue fluorescence. Hydrochloric acid added to the aqueous solution has no action ; caustic soda gives a milky precipitate which is soluble in ether with a steel blue fluorescence.—*Dyes* mordanted cotton yellow, silk yellow with a greenish fluorescence. Not now in use.

Flavaniline S—

(MEISTER, LUCIUS.)—Sodium salt of the sulphonic acid of Flavaniline. (1881.)— *Orange-yellow powder*, soluble in water to a yellow solution, in strong sulphuric acid to a colourless solution, which on diluting with water turns yellow. Hydrochloric acid added to the aqueous solution has little action ; caustic soda decolourises it.—*Dyes* wool yellow from acid baths. Not now in use.

Flavaurin—

The ammonium salt of dinitrophenol-*p*-monosulphonic acid

$$C_6H_2ONH_4(NO_2)_2SO_3NH_4$$

obtained by acting on mononitrophenol-*p*-sulphonic acid with nitric acid. (1883)— BEYER AND KEGEL.—*Reddish yellow powder*, slightly soluble in water. Hydrochloric acid turns the colour of the solution to a wine-red, but gives no precipitate. Dissolves in strong sulphuric acid to a wine-red solution.—*Dyes* wool and silk yellow in acid baths ; the shades are not fast. Known also as New Yellow (BEYER AND KEGEL).

Flavophenin—

(BADISCHE.)—Syn. of Chrysamine G. Not now sent out under this name.

Flavopurpurine—

Trihydroxyanthraquinone

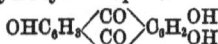

$$OHC_6H_2\diagup{CO}\atop{CO}\diagdown C_6H_2{OH}\atop{OH}$$

is found in commercial Alizarine ; it resembles Anthrapurpurin very closely. It crystallises in yellow needles, slightly soluble in water, readily soluble in alcohol. Melts above 330° C. Dissolves in caustic alkalies with a purple colour. Flavopurpurine dyes mordanted fibres very similar to Alizarine ; the reds are somewhat duller and yellower, the browns are not so bright. Is known as Alizarine YCA (BRITISH ALIZARINE COMPANY), Alizarine G (BADISCHE), Alizarine SDG (MEISTER), Alizarine X (BAYER), Alizarine FA (LEVERKUS). See ALIZARINE.

Fluorescein—

$$C_6H_4\diagup{CO.C_6H_3.OH.}\atop{COC_6H_3OH.}\diagdown O$$

is the base of the Eosin dyes. It is prepared by heating resorcin with phthalic anhydride to 200° C., and may therefore be called Resorcin Phthalein, and is analogous to phenol phthalein. It forms dark brown crystals, insoluble in water, but soluble in weak alkaline solutions to a red solution, which, when dilute, has a splendid yellowish green fluorescence. Soluble in alcohol, the solution has a green fluorescence.—*Dyes* wool and silk yellow, and the shades have a faint fluorescence, but it has little tinctorial value, its dyeing power being weak, and the shades fugitive.

Fluorescein Blue —

(SOCY. CHEM. IND., BASLE.)— $$C_{18}H_8Br_5N_2O_5NH_4$$

obtained by treating in potassium carbonate solution Diazoresorufin with Bromine, and precipitating with an acid. (1880.)— *Brown-red solution, containing green crystals*, soluble in water to a red-violet solution, having a green fluorescence. Strong sulphuric acid dissolves the dry dyestuff to a blue solution ; on diluting with water this turns first violet, then a brown-red precipitate is obtained. Hydrochloric acid gives a yellow-brown precipitate in the aqueous solution ; caustic soda has little action.—*Dyes* silk and wool from neutral baths blue with a brownish fluorescence. Also known as Resorcin Blue.

Formyl Violet S4B—

(CASSELLA AND CO.) — 1891. — *Purple powder*, with a bronze cast, soluble in water to a bright reddish violet solution, in alcohol soluble with a red-violet colour, in acetic acid to a blue solution, in strong sulphuric acid to a dark yellow solution, which on dilution with water turns first pale yellow, then on standing becomes colourless. Hydrochloric acid added to the aqueous solution turns it to pale greenish yellow ; on standing the solution becomes colourless. Caustic soda turns it a blue.— *Dyes* wool and silk from weakly acid baths a fine blue-violet, turned green by acids, bluer by alkalies, moderately fast to soaping, rather faster than acid violet.

Fuchsin—

German name of Magenta.

Fuchsin S—

Syn. of Acid Magenta.

Fuchsiacine—

So-called from the resemblance of the colour to that of the fuchsia, obtained by heating aniline and stannic chloride. It is a kind of impure magenta. *English Patent 921, 1859.*

G

Galamine Blue—

(GEIGY.) — Obtained by acting with gallaminic acid on nitroso-dimethyl aniline (1889.) *English Patent 2,941, 1889.—Dirty green paste,* slightly soluble in boiling water to a blue solution, in strong sulphuric acid to a red solution; on diluting with water the red colour remains; soluble in ammonia to a violet solution. Hydrochloric acid added to the aqueous solution reddens it; caustic soda turns it violet.—*Dyes* wool mordanted with chrome a dull reddish blue, fairly fast to light, not fast to soaping or acids. Useful in printing.

Gallein—

$$C \begin{cases} C_6H_2.OH.O \\ C_6H_2.OH.O \\ C_6H_4.CO.O \end{cases} \!\!\! > O$$

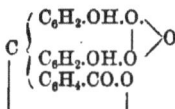

obtained by heating phthalic anhydride with gallic acid or pyrogallol. (1870.)— *Violet paste,* when dry dark green metallic lustrous crystalline powder, very slightly soluble in cold, easily in hot water to a scarlet-red solution, soluble in strong sulphuric acid to reddish yellow solution; on dilution with water there is no change. Caustic soda turns the colour of the aqueous solution a fine blue.—*Dyes* mordanted cotton violet, fairly fast. Also known as Alizarine Violet, Anthracene Violet.

Gallocyanine—

(DURAND AND HUGUENIN.) — Chloride of dimethylphenylammoniumdioxypheno-azin carbonic acid

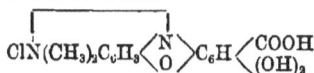

$$ClN(CH_3)_2C_6H_3 \langle {N \atop O} \rangle C_6H \langle {COOH \atop (OH)_2}$$

obtained by acting with nitrate of nitro-sodimethylanilino upon gallic acid or tannin. (1881.) (H. KOECHLIN).—*Greenish grey paste,* which when dried gives a *bronze-coloured* powder. Insoluble in water, soluble in alcohol to a blue-violet solution, in strong sulphuric acid to a pale bright blue solution; on diluting with water gives a bluish red solution; soluble in hydrochloric acid to a crimson-red solution, in caustic soda to a red-violet solution.—

Dyes chrome-mordanted wool blue-violet shades, fairly fast to acids, light and washing. Applied in calico printing with a chrome mordant. Known also as Solid Violet.

Galloflavine—

(BADISCHE.)—Prepared by oxidising gallic acid in alkaline, aqueous, or alcoholic solution with air. (1886.) *English Patent 6,413, 1886.—Greenish yellow paste,* insoluble in water, very slightly soluble in alcohol; the solution has a pale yellow colour and weak green fluorescence, soluble in strong sulphuric acid to a scarlet solution; on diluting with water a greyish precipitate is obtained. Hydrochloric acid added to the paste, diluted with water, causes little change; caustic soda turns it red.—*Dyes* wool and silk mordanted with chrome yellows, fast to light and soap. Applied in calico printing with a chrome mordant, give greenish yellow shades, fast to light and soaping.

Gambine B—

(READ HOLLIDAY AND SONS.) — Nitroso-dioxy-naphthaline

$$C_{10}H_5NO(OH)_2$$

(1890.) *English Patent 1,813, 1890.— Greenish black liquid,* slightly soluble in boiling water, soluble in alcohol to a dark olive solution, in acetic acid to a yellow-brown solution, in strong sulphuric acid to an olive-green solution; on diluting with water a reddish brown precipitate in a yellow solution is obtained. Hydrochloric acid added to the aqueous solution gives a brown precipitate; caustic soda gives a yellow solution.—*Dyes* wool and jute mordanted with bichrome olive-browns, with iron olive-greens, fast to acids, turned a little darker by alkalies, and fairly fast to light.

Gambine R—

(READ HOLLIDAY AND SONS.) — Beta-nitroso-alpha-naphthol

$$C_{10}H_6NOOH$$

(1889.) *English Patent 10,542, 1886.— Olive-yellow paste,* containing about 30 per cent. of solid colour, insoluble in cold, soluble in boiling water to an olive-yellow solution, in alcohol to a greenish olive solution, in strong sulphuric acid to a reddish brown solution, which on diluting with water turns yellow. Hydrochloric acid added to the aqueous solution has no action; caustic soda turns it greener.— *Dyes* wool mordanted with chrome deep-red brown, with iron olive-green, fast to light, air, acids and washing.

Gambine Y—

(READ HOLLIDAY AND SONS) — Alpha-nitroso-beta-naphthol

$$C_{10}H_6NOOH$$

(1889.) *English Patent* 10,542, 1886.— *Olive-yellow paste*, containing about 30 per cent. of solid colour, insoluble in cold, soluble in boiling water to an olive-yellow solution, soluble in alcohol to an olive-green solution, in strong sulphuric acid to a reddish brown solution, which on diluting with water turns yellow. Hydrochloric acid added to the aqueous solution has no action; caustic soda turn it greener. —*Dyes* wool mordanted with chrome a reddish brown, with iron deep olive-green, fast to acids, light, air and washing.

Gentiana Blue 6B—

(ACTIENGESELLSCHAFT.)—Syn. of Opal Blue.

Gentianine A—

(GEIGY.)—Dimethylderivative of Lauth's violet.—*Greenish bronze powder*, soluble in water to a bright blue solution, in alcohol and acetic acid to bright blue solutions. In strong sulphuric acid it dissolves with a dark green colour; on diluting the solution with water it turns a dark blue. Hydrochloric acid added to the aqueous solution has no action; caustic soda produces a dark violet precipitate.—*Dyes* tannin mordanted cotton a fine blue, fast to dilute acids; strong mineral acids turn it greenish, alkalies turn it violet, not fast to soaping or light.—*Dyes* wool and silk from neutral baths reddish blues, not fast to soap or light.

Girofle—

(DURAND AND HUGUENIN.) — Xylyldimethylamidophenyl-xylylazonium chloride

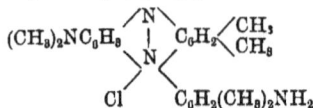

$$(CH_3)_2NC_6H_3 \underset{Cl}{\overset{N}{\underset{N}{\mid}}} C_6H_2 \underset{C_6H_2(CH_3)_2NH_2}{\overset{CH_3}{\underset{CH_3}{<}}}$$

acting with nitrate of nitrosodimethylaniline upon a mixture of nitrate of *m*-xylidine and *p*-xylidine. — *Brown paste or greyish green powder*, soluble in water to a crimson solution, in alcohol to a red solution, in strong sulphuric acid to a yellow-green solution; on diluting with water this turns first blue, then red. Hydrochloric acid gives a blue solution. Caustic soda gives, from strong solutions, a red precipitate.—*Dyes* cotton mordanted with tannin and antimony, wool and silk in neutral baths, red-violet.

Gladiolus Scarlet—

(WILLIAMS BROS. AND CO.)—(1888).— *Scarlet coloured paste*, insoluble in water, readily soluble in alkaline solutions; acids give in these alkaline solutions a dark purplish precipitate.—*Dyes* cotton from an alkaline bath fine brilliant scarlet, not fast to light and air, fast to soap and alkalies, turned purple by acids.

Gold Orange—

(BAYER.)—Syn. of Orange II.

Gold Yellow—

Syn. of Naphthol Yellow.

Gold Yellow—

(BAYER.)—Syn. of Resorcin Yellow.

Greenish Blue—

(MEISTER, LUCIUS AND BRUNING.)— Hydrochlorate of tri-*p*-tolyl-rosaniline. Properties similar to Spirit Blue.

Grenat—

A red colour obtained from the crude magenta melt after extracting the magenta.

Guernsey Blue—

One of the Induline blues.

Guinea Green B—

(ACTIENGESELLSCHAFT.)—Sodium salt of diethyl-dibenzyl-diamido-triphenyl carbinol disulphonic acid

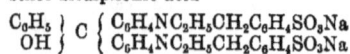

$$\left. \begin{array}{c} C_6H_5 \\ OH \end{array} \right\} C \left\{ \begin{array}{c} C_6H_4NC_2H_5CH_2C_6H_4SO_3Na \\ C_6H_4NC_2H_5CH_2C_6H_4SO_3Na \end{array} \right.$$

(1883.)—*Dark green powder*, soluble in water with a green colour, in strong sulphuric acid to a yellow solution, on diluting with water this turns first reddish, then yellow-green, finally green. Soluble in alcohol. Hydrochloric acid added to the aqueous solution turns it a brownish yellow; caustic soda gives a blackish green precipitate.—*Dyes* wool in acid baths, silk in a broken soap bath, fine bluish green, fairly fast to light, air and washing.

H

Harmaline—

A red colouring matter obtained from aniline, by acting with peroxide of manganese; it is therefore a more or less impure magenta. *English Patent* 1,155, 1859.

Hebburn Fast Blue—

(HEBBURN COLOUR CO.)—An improved make of Induline blue.

Helianthin—

(BADISCHE.)—Syn. of Orange III.

Hellochrysin—

(MEISTER, LUCIUS AND BRUNING.)—Sodium salt of tetranitro-*a*-naphthol

$$C_{10}H_3 \begin{array}{l} (NO_2)_4 \\ ONa \end{array}$$

obtained by treating tetranitro-*a*-bromnaphthalene with caustic soda. (1882.) (MERZ AND WEITH.)—*Dyes* wool and silk in acid bath fine gold-yellow shades, which are not fast to light. Not now used. Known also as Sun Yellow.

Helvetia Green—

Syn. of Acid Green (1).

Heliotrop—

(BAYER.)—

Tetrazodiphenetol $\left\{\begin{array}{l} \text{methyl-}B\text{-naphthylamine} \\ \text{sodium monosulphonate} \\ (delta) \\ \text{methyl-}B\text{-naphthylamine} \\ \text{sodium monosulphonate} \\ (delta) \end{array}\right.$

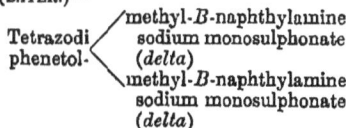

(1887). — *Dark purple - brown powder,* readily soluble in water to a crimson-red solution, in strong sulphuric acid to a blue solution ; on diluting with water a blue-violet precipitate is obtained. Hydrochloric acid throws down a violet precipitate from the aqueous solution; caustic soda has little action.—*Dyes* cotton from a soap bath reddish-violet shades, resembling the colour of the flower heliotrope, which are turned bluer by acids, slightly redder by alkalies, fairly fast to light.— *Dyes* wool from a salt bath bluish scarlet shades.

Hessian Purple B—

(LEONHARDT. BAYER. ACTIENGESELL-SCHAFT.)—

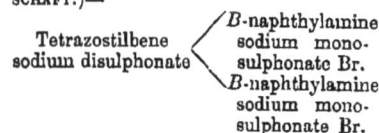

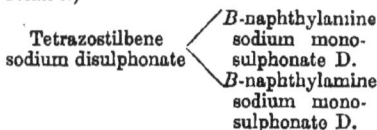

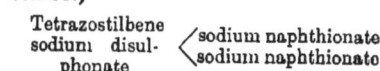

Tetrazostilbene sodium disulphonate $\left\{\begin{array}{l} B\text{-naphthylamine} \\ \text{sodium mono-} \\ \text{sulphonate Br.} \\ B\text{-naphthylamine} \\ \text{sodium mono-} \\ \text{sulphonate Br.} \end{array}\right.$

(1886.) *English Patent* 4,387, 1886.— *Brown-red powder,* completely soluble in water to a dark scarlet solution, in strong sulphuric acid to a violet solution ; on diluting with water a brown-black precipitate is obtained. Hydrochloric acid gives a bluish black precipitate, and caustic soda a crimson precipitate from the aqueous solution.—*Dyes* cotton, wool and silk in the same way as Hessian Purple P, but gives rather bluer shades of scarlet.

Hessian Purple D—

(LEONHARDT. BAYER. ACTIENGESELL-SCHAFT.)—

Tetrazostilbene sodium disulphonate $\left\{\begin{array}{l} B\text{-naphthylamine} \\ \text{sodium mono-} \\ \text{sulphonate D.} \\ B\text{-naphthylamine} \\ \text{sodium mono-} \\ \text{sulphonate D.} \end{array}\right.$

(1886.) *English Patent* 4,387, 1886.— *Blackish powder,* soluble in water to a red solution, in strong sulphuric acid to a violet solution; on diluting with water a brown solution is obtained. Hydrochloric acid added to the aqueous solution gives a brown precipitate; caustic soda turns the colour of the solution bluer.—*Dyes* cotton, wool and silk same way as Hessian Purple P, but the shades are rather bluer. The Hessian Purples P, B, and D are isomeric with each other.

Hessian Purple N—

(LEONHARDT. BAYER. ACTIENGESELL-SCHAFT.)—

Tetrazostilbene sodium disulphonate $\left\{\begin{array}{l} B\text{-naphthylamine} \\ B\text{-naphthylamine} \end{array}\right.$

(1886.)—*English Patent* 4,387, 1886.— *Brown-red powder,* soluble in water to a turbid red solution, in strong sulphuric acid to a blue solution ; on diluting with water a blue-black precipitate is obtained. Hydrochloric acid throws down a dark purple-brown precipitate from the aqueous solution. Caustic soda turns the colour of the solution slightly more scarlet.—*Dyes* cotton from a salt bath deep shades of bluish red ; not fast to light, exposure to air dulls them, acids turn them blue. Dyes wool from a salt bath redder shades than on cotton. Dyes silk from a neutral bath fine shades of scarlet.

Hessian Purple P—

(LEONHARDT. BAYER. ACTIENGESELL-SCHAFT.)—

Tetrazostilbene sodium disul-phonate $\left\{\begin{array}{l} \text{sodium naphthionate} \\ \text{sodium naphthionate} \end{array}\right.$

(1886.)—*English Patent* 4,387, 1886.— *Brown powder,* soluble in water to a scarlet solution, in strong sulphuric acid to a crimson solution ; on diluting with water a red-violet precipitate is obtained. Hydrochloric acid added to the aqueous solution gives a blue colouration ; caustic soda gives a bluish red solution.—*Dyes* cotton, wool, and silk similar shades to Hessian Purple N, and having the same properties.

Hessian Violet—

(LEONHARDT. BAYER. ACTIENGESELL-SCHAFT.)—

Tetrazostilbene ⟨a-naphthylamine
sodium disulphonate ⟨B-naphthol
(1886.) *English Patent* 4,387, 1886.—
Black powder, readily soluble in water to
a dark reddish violet solution, in strong
sulphuric acid to a blue solution; on diluting with water a violet precipitate is
obtained. Acids give in the aqueous
solution a dark blue precipitate, and
alkalies a red precipitate.—*Dyes* cotton
from a salt bath a dark blue-violet colour,
moderately fast to light, turned blue by
acids, reddened by alkalies, fast to washing. Dyes wool in a faintly acid bath dull
shades of brown.

Hessian Yellow—

(LEONHARDT. BAYER. ACTIENGESELL-SCHAFT.)—

Tetrazostilbene ⟨sodium salicylate
sodium disulphonate ⟨sodium salicylate
(1886.) *English Patent* 4,387, 1886.—
Yellow-brown powder, soluble in water to
a brownish yellow solution, in strong
sulphuric acid to a reddish violet solution;
on diluting with water a blackish brown
precipitate is obtained. Acids give in the
aqueous solution a dark greenish brown
precipitate, while alkalies turn the colour
of the solution scarlet.—*Dyes* cotton, wool,
and silk from neutral baths orange-yellows,
bright, fast to light, slightly dulled by
acids, reddened by alkalies, fast to washing.

Hofmann's Violet –

The composition of the products of
different makers is very variable; usually
it is the hydrochlorate of triethyl-rosaniline, but the acetate and chlorhydrate of
the mono, di, and tri ethyl or methyl-
rosaniline is often sold under different
marks

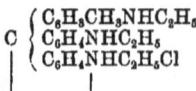

$$C \begin{cases} C_6H_5CH_3NHC_2H_5 \\ C_6H_4NHC_2H_5 \\ C_6H_4NHC_2H_5Cl \end{cases}$$

Hydrochlorate of triethyl-rosaniline, obtained by acting with methyl iodide or
methyl chloride, or ethyl bromide upon
magenta. 1863. (HOFMANN.) *English
Patent* 1,291, 1863.—The red shades of this
violet are usually in the form of a *green,
lustrous crystalline powder,* soluble in
water to a crimson solution, in strong sulphuric acid to a yellow-brown solution,
which is but little altered on diluting with
water. Hydrochloric acid added to the
aqueous solution turns it yellow-brown;
caustic soda gives a brown precipitate.—

Dyes wool in a neutral bath, silk also in a
neutral bath (cotton requires a tannin and
antimony mordant), fine bright bluish
reds, not fast to light or washing.
The blue shades of Hofmann's violets are
usually in the form of *greenish lustrous
pieces,* soluble in water to a blue-violet
coloured solution, in strong sulphuric acid
to a brownish yellow solution; on diluting
with water this turns first an olive-green,
then green, finally blue. Hydrochloric
acid gives first a green, then a yellow
colour to the aqueous solution; caustic soda
gives a brown-red precipitate.—*Dyes* the
fibres in the same way as the red shades, but
gives pure violet shades; not fast to light
or washing.
Known also as Iodine Violet, Dahlia,
Primula, Red Violet 5R, Extra Violet 5R,
Violet R.

I

Imperial Scarlet—

(BAYER.)—Syn. of Biebrich Scarlet.

Imperial Scarlet 3G—

Under this name was sent out a dyestuff made by combining B-naphthylamin
monosulphonic acid with B-naphthol, no
now sold.

Imperial Violet—

Syn. of Regina Purple.

Imperial Yellow—

Syn. of Aurantia.

Indamine—

(WM. NOETZEL AND CO.)—1890.—This
colouring matter is made in several brands,
GG, J, JO, R, 2R, 3R, and 6R, which dye
shades varying from a bright blue (GG)
to a crimson-red (6R). The three principal
brands are the following:—
INDAMINE GG—*A grey-black crystalline
powder,* soluble in water to a bright blue
solution, in alcohol to a dark blue, and in
acetic acid to bright blue solutions. In
strong sulphuric acid it dissolves with a
sage-green colour, which on diluting with
water turns to a bluish green. Hydrochloric acid added to the aqueous solution
has no action; caustic soda discharges
the colour, giving a precipitate of the colour
base.—*Dyes* tannin mordanted cotton a
fine bright blue, fast to acids, moderately
fast to alkalies and soaping, fast to light.
INDAMINE J — *Purple-red crystalline
powder,* soluble in water to a blue solution,
in alcohol and acetic acid to blue solutions,
in strong sulphuric acid to a bronze-green
solution, on diluting which with water a

turbid green liquid is obtained. Hydrochloric acid added to the aqueous solution has no action ; caustic soda throws down a colourless precipitate of the free base.— *Dyes* tannin mordanted cotton a fine indigo blue, fast to acids and light, moderately fast to soaping and alkalies.

INDAMINE 6R—*Greenish black powder*, soluble in water to a dark crimson solution in alcohol with the same colour. Soluble in strong sulphuric acid to a green solution ; on diluting with water a blue precipitate falls down. Hydrochloric acid added to the aqueous solution turns it blue. Caustic soda throws down a colourless precipitate of the free colour base.—*Dyes* tannin mordanted cotton fine deep bluish crimson. Acids turn the colour blue, alkalies darken the shade ; moderately fast to soaping, fast to light.

Indazine M—

(CASSELLA AND CO.)—Nitrosodimethylaniline diphenyl-*m*-phenylene diamine.— (See AZINES, page 14.) — *Blue - black powder*, soluble in water to a reddish blue solution, in strong sulphuric acid to a blue solution. Hydrochloric acid added to the aqueous solution turns it a purer blue ; caustic soda gives a reddish blue flocculent precipitate.—*Dyes* cotton mordanted with tannin and tartar emetic, wool and silk in slightly acid baths, fast to acids, light, and air.

Indian Yellow—

(BAYER.)—A mixture of nitro-diphenylamine orange and nitro-diphenylamine, obtained by treating diphenylamine orange with nitric acid. (1880.)—(KNECHT.)— *Ochre yellow powder*, very slightly soluble in cold water, more freely in hot water, in strong sulphuric acid to a crimson-red solution ; on diluting with water it turns first scarlet, then a yellow-brown precipitate falls down. Caustic soda turns the aqueous solution brown.—*Dyes* wool in acid baths. Also known as Azoacid Yellow (ACTIENGESELLSCHAFT), Azo-yellow (MEISTER, LUCIUS AND BRUNING), Azoflavin S (BADISCHE). This colouring matter is very similar to Curcumein.

Indophenine B—

(BAYER.)—One of the members of the Induline group. (1889.)—*Deep blue liquid*, slightly soluble in cold, easily in boiling water to a blue solution, in strong sulphuric acid to a violet solution ; on diluting with water this turns blue. Hydrochloric acid has no action on the aqueous solution. Caustic soda gives a purple precipitate.— *Dyes* cotton mordanted with tannin and

antimony blue, but the shades are poor. Applied in calico printing with good results. The colour is fast to acids, light, and washing.

Indophenol—

(DURAND AND HUGUENIN. CASSELLA AND CO.)—Produced by oxidising dimethyl-*para*-amidophenyl-*a*-oxy-*a*-naphthylamine

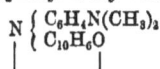

$$N \begin{cases} C_6H_4N(CH_3)_2 \\ C_{10}H_6O \end{cases}$$

obtained by acting with nitrosodimethylaniline upon *a*-naphthol, or by oxidising amido-dimethylaniline and *a*-naphthol. (1881.)—(KOECHLIN AND WITT.)—*Dark brown powder*, insoluble in water, soluble in alcohol with a blue colour, in strong sulphuric acid to a yellow-brown solution ; on diluting with water a brown precipitate is obtained. Hydrochloric acid added to the alcoholic solution gives a red-brown solution ; on adding zinc powder the colour is discharged ; the colour is restored on exposure to air or adding alkalies —*Dyes* fibres blue in the same way as indigo ; the shades are fairly fast to light, acids and washing. Now used in combination with indigo in one vat. (*English Patent* 10,496, 1888.)

Indophenol White—

(DURAND AND HUGUENIN. CASSELLA AND CO.)—

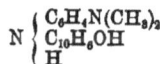

$$N \begin{cases} C_6H_4N(CH_3)_2 \\ C_{10}H_6OH \\ H \end{cases}$$

reduction product of indophenol made by treating with acetate of tin. (1881.) (KOECHLIN AND WITT.)—*Yellowish white paste*, soluble in water ; on adding ammonia or caustic soda, and agitating with air, a blue precipitate of indophenol is obtained.—*Dyes* fibres by immersion in an alkaline solution and subsequent exposure to air. (*See* INDOPHENOL.)

Induline—

Made in many brands and shades.

Sodium salt of the sulphonic acids of the spirit soluble indulines. (1867.)— COUPIER. *English Patent* 3,657, 1867.— *Brownish lustrous* (INDULINE) or *black lustrous* (NIGROSINE) *powder*, soluble in water to a blue-violet solution, in alcohol to a blue solution, in strong sulphuric acid to a blue solution ; on diluting with water a blue-violet solution is obtained ; acids have no action on the aqueous solution ; caustic soda gives a brown-violet precipitate.—*Dyes* wool and silk from faintly acid baths : induline a pure blue, nigrosine

H

a grey; very fast to acids, light and washing. Cotton can be dyed on a tannin mordant. Known also as Fast Blue R, Nigrosine Fast Blue B, Induline 3B, Induline 6B, Fast Blue green shade.

The exact chemical composition and constitution of the indulines and nigrosines has not yet been worked out. Many of them are not definite chemical compounds, but admixtures of various members of the Induline series.

Induline Spirit Soluble—

(ROBERTS, DALE, AND CO.)—

$$C_{18}H_{16}N_3Cl$$

obtained by heating (1) amidoazobenzol with aniline salt. (2) Nitrobenzol with aniline, aniline hydrochlorate and iron to 180° C. (3) Nitrophenol with aniline and aniline hydrochlorate. — (1866.) — CARO. The bluer shades 3B and 6B are got by using larger quantities of aniline ; by varying the proportions of the ingredients various shades are obtained.—*Blue-black powder*, insoluble in water, soluble in alcohol to a blue-violet solution, in strong sulphuric acid to a blue solution ; on diluting with water a blue precipitate is obtained. Soluble in acetic acid, acetin. Rarely used in dyeing ; used for colouring spirit varnishes. Used in calico printing in the form of solutions in acetic acid, acetin, gives fine blues, very fast to acids and washing. Also known as Fast Blue R (spirit soluble), Printing Blue, Acetin Blue, Indigene D and F, Nigrosine (spirit soluble), Coupier's Blue, Azodiphenyl Blue, Violaniline, Induline 3B (spirit soluble), Induline 6B (spirit soluble), Fast Blue B (spirit soluble).

Iodine Eosine B—

Syn. of Erythrosine.

Iodine Eosine G—

Syn. of Erythrosine G.

Iodine Green—

(READ HOLLIDAY AND SONS.) — Zinc chloride double salt of chlormethyl-hexamethylrosaniline hydrochlorate

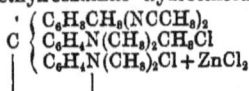

$$C\begin{cases} C_6H_5CH_2(NCCH_3)_2 \\ C_6H_4N(CH_3)_2CH_2Cl \\ C_6H_4N(CH_3)_2Cl + ZnCl_2 \end{cases}$$

obtained by acting with methyliodide or methylchloride upon rosaniline. (1866.) *English Patent* No. 1,340, 1866.— *Dark green pieces*, rather hard, easily

soluble in water to a blue-green solution, in strong sulphuric acid to a reddish yellow solution; if the colour has been prepared with methyliodide, on warming vapours of iodine are given off. This solution on diluting with water turns yellow-green. Paper or fibre stained with solutions of this colour on being dried and heated turn violet. Hydrochloric acid added to the aqueous solution turns it reddish yellow ; caustic soda decolorises it.—*Dyes* wool from a neutral bath, silk from an old soap bath; cotton requires a mordant of tannin and tartar emetic. Bluish shades of green are obtained, which are not fast to light or washing. Also known as Night Green.

Iodine Violet—

Syn. of Hofman's Violet.

Isatin Yellow—

(ACTIENGESELLSCHAFT.)—Sodium salt of phenyl-*p*-sulphonic acid oxazon isatin

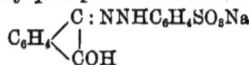

$$C_6H_4\begin{cases} C : NNHC_6H_4SO_3Na \\ COH \end{cases}$$

obtained by treating phenylhydrazin-*p*-sulphonic acid on isatin. (1886.)—*Orange-yellow powder*, soluble in water to a yellow solution, in strong sulphuric acid to a yellow-brown solution; on diluting with water this turns pale yellow. Hydrochloric acid added to the aqueous solution has little action ; caustic soda turns it browner.—*Dyes* wool from an acid bath greenish yellow, fast to acids and washing. Dyes silk from an acid bath bright greenish yellow.

J

Jet Black G—

(BAYER.)—Benzene sodium disulphonate azo-naphthalene-azo-benzyl napthylamine

$$(SO_3Na)_2C_6H_3N : NC_{10}H_6N : C_{10}H_5NHC_7H_7$$

obtained by diazotising benzene sodium disulphonate-azo-napthylamine and combining with benzyl naphthylamine. (1889.) —*Dyes* wool and silk from a salt bath greenish blue-black, fast to light, air, washing, and acids.

Jet Black R—

(BAYER.)—Isomeric with Jet Black G. (1889.)—*Dyes* wool and silk from a salt bath blue-black, fast to washing, alkalies, and acids.

L

Lacmoid--

Syn. of Resorcin Blue.

Lancaster Yellow—

Picramic acid azo-phenol

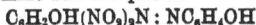

$$C_6H_2OH(NO_2)_2N : NC_6H_4OH$$

obtained by diazotising picramic acid and combining with phenol. (1877.)—GRIESS. *English Patent* 3,696, 1877.—*Blackish brown crystals*, very slightly soluble in water, soluble in strong sulphuric acid to a reddish yellow solution; caustic soda turns the colour of the aqueous solution to a yellowish red.—*Dyes* wool and silk in acid baths brownish yellow. Not now in use.

Lauth's Violet—

Hydrochlorate of Thionine

$$N \left\{ \begin{array}{l} C_6H_3NH_2 \cdot \\ C_6H_3NH_2Cl \cdot \end{array} \right\} S$$

obtained by the oxidation of *p*-phenylene diamine in sulphuretted hydrogen solution with ferric chloride. (1876.)—CH. LAUTH.— *Black-green powder with a metallic lustre*, slightly soluble in cold, easily in hot water, to a violet solution, in strong sulphuric acid to a yellow-green solution; on diluting with water this turns first blue, then violet. Hydrochloric acid turns the colour of the aqueous solution bluer; caustic soda gives a brown-red precipitate. Known also as Thionine.

Light Green—

Syn. of Methyl Green.

Light Green S—

(BADISCHE.)—Sodium salt of dimethyl-dibenzyl-diamido-triphenyl carbinol sulphonic acid. See ACID GREEN (2.) (1879.)

London Orange—

Syn. of Chrysoidine.

Luteolin—

Sodium salt of *m*-xylene sulphonic acid-azo-diphenylamine

$$C_6H_2(CH_3)_2SO_3NaN : NC_6H_4NHC_6H_5$$

obtained by diazotising *m*-xylidine monosulphonic acid and combining with diphenylamine. (1883.)—*Orange-yellow powder*, slightly soluble in water to an orange solution, in strong sulphuric acid to a yellow-green solution ; on diluting with water a violet precipitate is obtained. Addition of hydrochloric acid to the aqueous solution throws down a violet precipitate ; caustic soda produces an orange-yellow precipitate.—*Dyes* wool in acid baths orange. Not now in use.

Lutecienne—

A mixture of Orange II. and Fast Red A.

M

Magdala Red—

(DURAND AND HUGUENIN.)—Diamidonaphthyl naphthazonium chloride

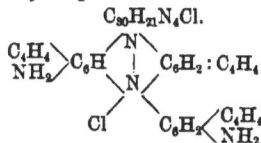

$$C_{30}H_{21}N_4Cl.$$

$$\begin{array}{c} C_4H_4 \\ NH_2 \end{array} \rangle C_6H \begin{array}{c} N \\ | \\ N \end{array} C_6H_2 : C_4H_4$$

$$Cl \qquad C_6H_2 \langle \begin{array}{c} C_4H_4 \\ NH_2 \end{array}$$

obtained by heating *a*-amido-azo-naphthene with *a*-naphthylamine. (1868.)—*Dark brown powder*, slightly soluble in boiling water, in strong sulphuric acid with a grey-violet colour ; on diluting with water this gives a violet-red precipitate. Hydrochloric acid added to the aqueous solution gives a violet colour ; caustic soda a red-violet precipitate.—*Dyes* cotton mordanted with tannin and tartar emetic rose-red shades. Dyes silk from an acidulated soap bath fine rose-reds having a weak fluorescence.

Magenta—

This colouring matter is usually the hydrochlorate or acetate of the base rosaniline, tolyl-diphenyl-triamido carbinol

$$C \left\{ \begin{array}{l} C_6H_3CH_3 \cdot NH_2 \\ C_6H_4NH_2 \\ C_6H_4 \\ NH_2Cl \end{array} \right.$$

Occasionally the sulphate and oxalate are met with. It is prepared by three processes:—1st, by oxidation of a mixture of aniline with *ortho* and *para* toluidine by means of arsenic acid. MEDLOCK, *English Patent*, 18th January, 1860; NICHOLSON, *English Patent*, 26th January, 1860; GIRARD AND DE LAIRE, *French Patent*, 26th May, 1860. 2nd, by oxidation of a mixture of aniline, *ortho* and *para* toluidine with nitro-benzol and nitro toluol. LAURENT AND CASTELHAZ, *French Patent*, 10th Dec., 1861; READ HOLLIDAY AND SONS, *English Patent* No. 2,564, 1865; COUPIER, 1869. 3rd, oxidation of a mixture of aniline and toluidine with mercuric nitrate or mercuric chloride. GERBER AND KELLER, *French Patent*, 29th Oct., 1859. It was described in 1856 by Natanson, who obtained it by treating aniline with ethylene chloride. Hofman prepared it in 1858 from carbon tetrachloride and aniline. Verguin in 1859 obtained it by treating aniline and toluidine with zinc chloride.—Magenta is met with in *large amorphous pieces or in fine diamond-shaped crystals*, this latter being the purer form. *It has a brilliant bronze-green lustre ;* it is soluble in water or alcohol to a deep bluish red solution, in strong sulphuric acid to a yellow-brown solution ; on diluting with water becomes almost colourless. Hydrochloric acid turns the aqueous solution yellow; caustic soda discharges the colour with precipitate of the free colour base rosaniline ; on adding acids the colour is restored.—*Dyes* wool and silk from neutral baths. Cotton requires mordanting with tannin and tartar emetic. Fine blue-red shades are obtained, which are not fast to light or washing. Known also as Fuchsin, Rubin, Aniline Red, Rosein. (This is generally applied to Acetate Magenta). The *para*-rosaniline magenta has a bluer shade than the *ortho*-rosaniline magenta.

In the purification of the crude magenta melt a variety of impurities of a red colour are separated out ; these after more or less purifying are sold as Maroon, Grenat, Cerise, Geranium, Azalein, Crimson, Solferino, &c.

Maize—

Syn. of Curcumine S.

Malachite Green—

(ACTIENGESELLSCHAFT.)—This colouring matter varies slightly in composition, as it is issued by different makers ; it is either the zinc or iron double salt, or the oxalate of tetra-methyl di-*p*-amidotriphenyl carbinol

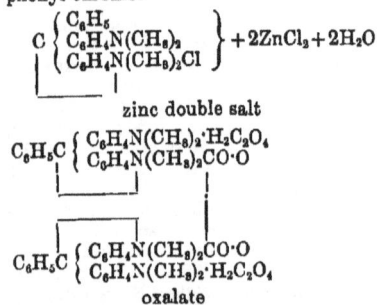

$$C \begin{cases} C_6H_5 \\ C_6H_4N(CH_3)_2 \\ C_6H_4N(CH_3)_2Cl \end{cases} + 2ZnCl_2 + 2H_2O$$

zinc double salt

$$C_6H_5C \begin{cases} C_6H_4N(CH_3)_2 \cdot H_2C_2O_4 \\ C_6H_4N(CH_3)_2CO \cdot O \end{cases}$$

$$C_6H_5C \begin{cases} C_6H_4N(CH_3)_2CO \cdot O \\ C_6H_4N(CH_3)_2 \cdot H_2C_2O_4 \end{cases}$$

oxalate

obtained by acting with benzotrichloride upon dimethylaniline in contact with chloride of zinc. This is not now in use. Also by acting with benzaldehyde upon dimethylaniline, and oxidising the tetramethyl-*p*-amido triphenylmethane so produced. (1878.)—*The zinc double salt is a bright yellow prismatic crystal; the oxalate is in the form of green plates having a metallic lustre.* Soluble in water to a bluish green solution, in strong sulphuric acid to a yellow solution; on diluting with water this turns first dark yellow, then a yellow-green, finally after adding much water a green. Soluble in alcohol and in amyl-alcohol. Addition of hydrochloric acid to the aqueous solution turns it reddish-yellow ; caustic soda throws down the base as a leaf-green precipitate, soluble in ether.—*Dyes* wool from a neutral bath bluish green, silk from a neutral bath blue-green; cotton requires a tannin mordant, leather and jute are dyed direct. The shades are fine blue-green, moderately fast to light and washing, not fast to acids. Also known as Victoria Green (BADISCHE), New Green (BAYER); Solid Green (CASSELLA), Diamond Green, Fast Green, Benzal Green.

Malachite Green Spirit Soluble—

The picrate of tetramethyl-di-*p*-amido triphenyl carbinol. (See MALACHITE GREEN.) Used to a limited extent for colouring varnishes. Is insoluble in water, soluble in alcohol.

Malta Grey—

Syn. of Nigrisin.

Manchester Brown—

Syn. of Bismarck Brown.

Manchester Yellow—

Syn. of Naphthol Yellow.

Mandarin—

Syn. of Orange II.

Mandarin G Extra—

(ACTIENGESELLSCHAFT). — Syn. of Orange II.

Mandarin GR—

(ACTIENGESELLSCHAFT). — Syn. of Orange I.

Maroon—

A colouring matter obtained from crud magenta.

Martius Yellow—

Syn. of Naphthol Yellow.

Mauve Dye—

An old name for Mauvein.

Mauveine—

(PERKIN AND SON.)—

$$C_{27}H_{25}N_4Cl$$

obtained by the oxidation of toluidine and aniline. (1856.) PERKIN. *English Patent* No. 1,984, 1856. Proceedings of Royal Society, 35, 717.—*Red violet paste*, insoluble in cold, slightly soluble in hot water to a violet-red solution, in strong sulphuric acid to an olive-green solution; on diluting with water this turns first green, then blue, finally red-violet. Acids have no action in the aqueous solution; caustic soda gives a blue-violet precipitate. Very little used now.—*Dyes* silk direct a reddish violet. Cotton requires a mordant of tannin and tartar emetic. Also known as Rosolane (POIRRIER); Chrome Violet, Mauve, Aniline Purple, Indisin, Perkin's Violet, Aniline Violet, Violein, Purpurin.

Meldola's Blue—

(BROOKE, SIMPSON, AND SPILLER.)— Chloride of dimethylphenylammonium-B-naphthoxazine.

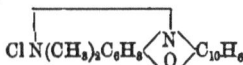

$$Cl\ N(CH_3)_2C_6H_3\diagdown\diagup^N_O\diagup C_{10}H_6$$

obtained by acting with nitrate of nitrosodimethylaniline upon *B* · naphthol. (1879.) R. MELDOLA.—*Dark violet bronzy powder*, easily soluble in water to a blue-violet solution, in alcohol to a blue solution, in strong sulphuric acid to a dark green solution; on diluting with water this turns blue. Hydrochloric acid turns the colour of the aqueous solution blue; caustic soda gives a brown precipitate.— *Dyes* cotton mordanted with tannin and tartar emetic indigo blue shades, very fast to light, acids and washing. Also known as New Blue (CASSELLA), Naphthylene Blue R, Crystal (BAYER), Fast Cotton Blue 2B, Fast Cotton Blue R Crystal (ACTIEN-GESELLSCHAFT), Cotton Blue R (BADISCHE), Meldoline Blue. There are several shades of this blue.

The New Blue R of Messrs. Cassella and Co. is, when pure, the chloride of dimethyl-amido-naphthophenoxazine.

Metamine Blue—

(LEONHARDT.)—(1889.)—*Blackish powder*, somewhat crystalline, soluble in water to a violet solution, in alcohol to a dark reddish blue solution, in acetic acid to a dark blue solution, in strong sulphuric acid to a deep olive-green solution, from which, on diluting with water, a dark violet-blue precipitate falls down. Hydrochloric acid added to the aqueous solution throws down a dark violet-blue precipitate. Caustic soda decolorises the aqueous

solutions.—*Dyes* cotton which has been mordanted with tannin, wool and silk from neutral baths, deep blue of a slight reddish hue. Fast to acids, soaping, and moderately fast to light.

Metaphenylene Blue B.—

(CASSELLA AND CO.)—Nitrosodimethyl-aniline + Di-o-tolyl-m-phenylene diamine. (1890.)—*Bronzy blue powder*, soluble in water to a dark violet-blue solution, in alcohol and acetic acid to blue solutions, in strong sulphuric acid to a violet-brown solution, which on diluting with water turns to a blackish green. Hydrochloric acid added to the aqueous solution turns it darker, while caustic soda throws down a violet precipitate.—*Dyes* tannin mordanted cotton, wool and silk from neutral baths deep blue of a reddish tone. Turned blue-green by acids, redder by alkalies; not quite fast to soaping, moderately fast to light.

Metaphenylene Blue R has very similar properties.

Methyldiphenylamine Blue—

Hydrochlorate of dimethyl-triphenyl-triamido-phenyl carbinol

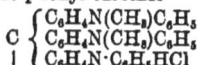

$$C\begin{cases}C_6H_4N(CH_3)C_6H_5\\C_6H_4N(CH_3)C_6H_5\\C_6H_4N\cdot C_6H_5HCl\end{cases}$$

obtained by acting upon methyl-diphenyl-amine with oxalic acid, copper nitrate, or chloranil. (1874.) GIRARD. *English Patent* No. 2,847, 4th July, 1874.— *Copper lustrous pieces*, insoluble in water, soluble in alcohol to a blue solution, in strong sulphuric acid to a brown-yellow solution; on diluting with water a blue precipitate is obtained. Caustic soda added to the alcohol solution turns it red. Acids do not change it.— *Dyes* wool and silk red shades of blue. Not now in use.

Methyl Blue—

(CASSELLA.)—Sodium salt of triphenyl-para-rosaniline trisulphonic acid, a pure form of the Soluble Blue (which see).

Methylene Blue—

(BADISCHE.)—Either the hydrochlorate of tetramethylthionine or the zinc double salt

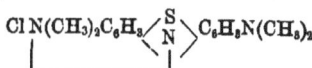

$$Cl\ N(CH_3)_2C_6H_3\diagdown^S_N\diagup C_6H_3N(CH_3)_2$$

is obtained in various ways. 1. Treating an acid solution of nitrosodimethylaniline with sulphuretted hydrogen, when the leuco base of the methylene blue is formed; this is then oxidised. (1876.)— CARO. *English Patent* 3,751, 1877.— 2. Nitrosodimethylaniline is dissolved in sulphuric acid, and treated with sulphide of zinc, when the leuco base of the blue is formed. (1882.) — OEHLER. (Ethylene Blue.) It can also be obtained from thionine sulphonic acid

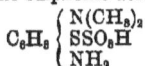

$$C_6H_3 \begin{cases} N(CH_3)_2 \\ SSO_3H \\ NH_2 \end{cases}$$

by several processes.—*Dark blue lustrous powder*, soluble in water to a bright blue solution, in alcohol and in strong sulphuric acid to a yellow-green solution; on diluting with water this turns blue. Hydrochloric acid added to the aqueous solution causes little change; dilute caustic soda turns it violet; strong caustic soda gives a violet precipitate.—*Dyes* cotton mordanted with tannin and tartar emetic, wool and silk in neutral baths, fine bright shades of greenish blue, fairly fast to light, acids, and washing. Known also as Ethylene Blue.

Methyl Eosin—
(ACTIENGESELLSCHAFT.)—Syn. of Eosin BN.

Methyl Eosin—
(MONNET. SOCY. CHEM. IND., BASLE).— Syn. of Erythrin.

Methyl Fluorescein—
is sold as a colouring matter under the name of Uranine.

Methyl Green—
(BAYER.)—Zinc chloride double salt of the hydrochlorate of chormethyl-hexamethyl para-rosaniline

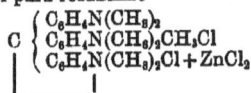

$$C \begin{cases} C_6H_4N(CH_3)_2 \\ C_6H_4N(CH_3)_2CH_3Cl \\ C_6H_4N(CH_3)_2Cl + ZnCl_2 \end{cases}$$

obtained either by acting with methyl nitrate upon methyl violet, or by acting with methyl chloride upon methyl violet in amyl-alcohol solution. (1871.)—*Green crystals*, easily soluble in water to a blue-green solution, in strong sulphuric acid to a reddish yellow solution; on diluting with water turns yellow-green. Soluble in alcohol, insoluble in amyl-alcohol. On adding hydrochloric acid to the aqueous solution it turns reddish yellow; caustic soda discharges the colour. Fibres coloured with methyl green on being heated turn violet.—*Dyes* wool in neutral bath, silk in old boiled off liquor, cotton mordanted with tannin fine green shades

with a slight blue tint, not fast to light, washing, or heat. Also known as Paris Green, Light Green, Etincelle Green, Powder Green.

Methyl Orange—
Syn. of Orange III.

Methyl Purple—
(BADISCHE.)—Hexamethyl para-rosaniline hydrochlorate

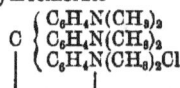

$$C \begin{cases} C_6H_4N(CH_3)_2 \\ C_6H_4N(CH_3)_2 \\ C_6H_4N(CH_3)_2Cl \end{cases}$$

obtained by treating dimethylaniline with phosgene. (1883.) *English Patents*, 4,428, 1883; 11,030, 1884; 5,038, 1884.

Methyl Violet B—
Hydrochlorate of pentamethyl-para-rosaniline

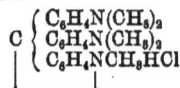

$$C \begin{cases} C_6H_4N(CH_3)_2 \\ C_6H_4N(CH_3)_2 \\ C_6H_4NCH_3HCl \end{cases}$$

obtained by oxidising dimethylaniline with copper chloride. (1861.)—LAUTH.— *Greenish lustrous masses* or *powder*, soluble in water or alcohol to a violet solution, in strong sulphuric acid to a yellow solution; on diluting with water this turns first yellow-green, then green-blue, finally violet. Hydrochloric acid turns the aqueous solution first green, then yellow-brown; caustic soda gives a brown-red precipitate.—*Dyes* wool and silk from neutral baths; cotton requires mordanting with tannin and tartar emetic. Fine bright shades of .violet are obtained, which are not fast to light or washing. Known also as Paris Violet.

Methyl Violet 6B—
Syn. of Benzyl Violet.

Methyl Water Blue—
(BADISCHE.) — Syn. of Methyl Blue (CASSELLA).

Metanil Orange I--
(SOCY. CHEM. IND., BASLE.) — Sodium salt of *m*-amido-benzene sulphonic acid-azo-*a*-naphthol

$$C_6H_4SO_3NaN : NC_{10}H_6OH$$

obtained by diazotising *m*-amido-benzene sulphonic acid, and combining with *a*-naphthol.

Metanil Orange II—
(SOCY. CHEM. IND., BASLE.)—Sodium salt

of m-amido-benzene sulphonic acid-azo-B-naphthol

$$C_6H_4SO_3NaN : NC_{10}H_6OH$$

obtained by diazotising m-amido-benzene sulphonic acid, and combining with B-naphthol.

Metanil Yellow—

(OEHLER. BADISCHE. ACTIENGESELLSCHAFT.) — Sodium salt of m-amido-benzene sulphonic acid - azo - diphenyl - amine

$$C_6H_4SO_3Na N : NC_6H_4NHC_6H_5$$

obtained by diazotising m-amido-benzene sulphonic acid and combining with diphenylamine. (1882.) — *Brown-yellow powder*, soluble in water with an orange-yellow colour, in strong sulphuric acid to a violet solution; on diluting with water the colour turns a crimson-red. On adding hydrochloric acid to the aqueous solution a precipitate in a crimson solution is obtained; caustic soda causes no change in the aqueous solution. — *Dyes* wool and silk in acid baths an orange-yellow. Also known as Orange M N (SOCY. CHEM. IND., BASLE).

Metanil Yellow S—

Sulphonic acid of Metanil Yellow. Properties same as Metanil Yellow.

Methods of Dyeing with the Coal Tar Colours—

To avoid a good deal of repetition, and to make the Dictionary as valuable a guide as possible to the dyer in the application of the coal tar colours to the fibres, general outlines of the most improved methods are here given.

COTTON DYEING.

METHOD A—The cotton is wetted out and worked in a bath of 1 per cent. for pale shades to 3 per cent. for deep shades, of tannic acid. For dark colours, such as Magenta, Clarets, New Blues, Blacks, sumac extract from 5 to 15 per cent. may be substituted. The bath is heated to from 150deg. to 170deg. F., the cotton entered and worked for half an hour, then allowed to steep for six hours, or better overnight in the case of yarns; piece goods are run slowly through the warm tannin bath, and allowed to lie overnight. The yarn is wrung out, and then entered into a bath of tartar emetic, containing from ½ to 2½ per cent., according to the depth of tint to be dyed. Piece goods are passed at once into the tartar emetic bath. The fabrics are then washed, and are then entered into the dye-bath, in which they are treated at from 180deg. to 190deg. F.,

until the full shade required is obtained. The goods should be dried after passing through the tartar emetic bath. The following colouring matters are dyed in this way. All basic aniline colours, Magenta, Safranine, Bismarck Browns, Chrysoidine, Violets (Methyl, Paris, Hofmann, Benzyl), Greens (Malachite, Iodine, Methyl, Victoria, New, Azine), Auramine, Phosphine, Rhodamine, Blues, New Blues, Indazine, Naphthylene Blue, Nile Blue, Thioflavine I, Benzoflavine, etc. These colouring matters can be mixed together for the production of compound shades.

METHOD B—The cotton is prepared by steeping in a bath of 10 per cent. of soap or Turkey red oil, wrung out, these operations being repeated several times; then it is dried slowly in a stove, passed through a cold bath of acetate of alumina at 8deg. Tw., wrung out, passed through a weak soap bath (1lb. of soap in 10 gallons of water), rinsed, and finally dyed in a fresh bath containing the colouring matter only. Applicable to Alizarine colours, Azure Blues, Crocein Scarlets, and the azo colours generally, but in these cases the results are not satisfactory.

METHOD C—The cotton is mordanted as in Method A, and 2 per cent. of alum is added to the dye-bath; for pale shades it is simply necessary to add 5 per cent. of alum to the dye-bath and dye at the boil. Applicable to Cotton Blues, Crocein Scarlets, Biebrich Scarlet, Cotton Scarlets, Eosines.

METHOD D—The cotton is entered into a cold bath of stannate of soda 6 to 8deg. Tw. for one hour, or best left overnight, wrung, and then run for 30 minutes through a cold bath of 15 to 20 per cent. of alum, or 7½ to 10 per cent. of sulphate of alumina. The cotton is wrung, rinsed, and then dyed at a gentle heat. Applicable to Crocein Scarlets, Cotton Scarlets, Biebrich Scarlet, Eosines, most azo colours, but the shades so got are not fast to soaping.

METHOD E—A bath containing from 10 to 20 per cent. of common salt, and the colour is made; the goods are entered about 150deg. F., and the temperature raised to the boil, the dyeing operation taking about one hour; the goods are then rinsed and dried. Applicable to all Benzidin colours, Hessian colours, Titan colours, Benzo-Browns, Benzo-Black-Blue, Jet Black, Cotton Yellow, Salmon Red, Diamine colours, Thioflavine S, Turmeric, Primuline, Auroline, Carnotine, Clayton Yellow, Erika, Sapphire Blue. These colours may be mixed together to form compound shades.

METHOD F—The dye-bath is made with 10 per cent. of soda or potash, borax, phosphate of soda, or sulphate of soda, and 2½ per cent. of soap. The dyeing is done as in the last. Congo Reds, Benzopurpurines, Roseazurines, Diamine Reds, Brahma Reds, Naphthylene Reds, Diamine Scarlet B, are dyed with potash or soda, although either of the other salts may be used. Chrysamine, Chrysophenine, Diamine Yellow N, Thioflavine S, are best dyed with phosphate of soda. The Blues of the Benizidine colours—Azo Blue, Benzoazurines, Brilliant Azurine, Sulphon Azurine, Diamine Blue, Diamine Black, Azo Violet, Diamine Violet—are best dyed with Glauber's salt or borax. Any of these colours may be mixed together to form compound shades.

WOOL DYEING.

METHOD A—The dye-bath is made with from 2 to 10 per cent. of Glauber's salt, and the colouring matter. The dyeing is done at the boil. Applicable to all the dyestuffs named in Method A, Cotton Dyeing.

METHOD B—The dye-bath is made with 2 to 10 per cent. of Glauber's salt, ½ to 2 per cent. of sulphuric acid, and the colouring matter. The goods are in both this and the previous method best entered at about 150deg. F., and after working a few minutes at this, raising to the boil and working until the right shade is obtained. Applicable to all so-called acid or azo colours. Scarlets, Oranges, Acid Greens, Acid Magenta, Acid Browns, Acid Violets, Croceines, Navy Blues, Cotton Blues, Soluble Blues, Indulines, Nigrosines, Ponceaus, Fast Reds, Cloth Reds, Naphthol Black, Naphthol Green.

METHOD C—The colouring matter is dissolved with borax or soda crystals; the wool is boiled with this for thirty minutes to one hour, then the colour is developed by passing through a bath of weak sulphuric acid. Applicable to Alkali or Nicholson Blues, Alkali Violet, Alkali Green.

METHOD D—The dye-bath is made with Glauber's salt 10 per cent., acetic acid 2 per cent., and the colouring matter, the process being carried on at the boil. Eosines, Erythrosine, Phloxine, Rose Bengal, Rhodamines are thus dyed.

METHOD E—The dye-bath is made with from 10 to 20 per cent. of common salt, 1 to 2 per cent. acetic acid, and the colouring matter; the dyeing is done at the boil. Applicable to all the Benzidine colours, such as Benzopurpurines, Roseazurines, Deltapurpurines, Benzo-

azurines, Chrysamine, Diamine Reds, Hessian Reds, Chrysophenine, Titan Pink, Titan Red, Titan Yellow, Titan Scarlet, Jet Black, many of the basic aniline dyes, etc.

METHOD F—The goods are mordanted by boiling in a bath of 3 per cent. bichromate of potash, or 5 per cent. ferrous sulphate, or 5 per cent. of sulphate of alumina, with the addition of 1 to 2 per cent. sulphuric acid, or of 2 to 3 per cent. oxalic acid, or 3 per cent. tartar, the mordanting being done at the boil for one hour and a half. Which of these mordants is used depends upon the colouring matter used and the shade to be dyed. After mordanting the goods are rinsed, and then entered into the dye-bath at about 80deg. to 100deg F., and worked for half an hour; then the temperature is slowly raised to the boil, and the operation is continued for one hour and a half at that heat. Applicable to all so-called adjective colouring matters, the Alizarine colours, Anthracene Brown, Anthracene Yellow, Galloflavine, Gambines, Gallein, Gallocyanine, Gallacetophenone, Cœruleine, Dioxine, Cloth Reds, Cloth Orange, Diamond Black, Azo Green, etc.

SILK DYEING.

METHOD A—The dye-bath is made with ½ to 1½ per cent. of soap, or of old boiled off liquor, and the colouring matter. The dyeing is done at from 140deg. to 150deg. F. Applicable to all basic colours, as in Cotton Dyeing, Method A.

METHOD B—The dye-bath is made with ½ to 1½ per cent. of soap, or a quantity of old boiled off liquor, a little acid, ¼ to ½ per cent., and the colouring matter; the dyeing being done at from 170deg. to 212deg. F. In both A and B the soap may be omitted with advantage in some cases. Applicable to all acid and azo colours named in Wool Dyeing, Method B.

METHOD C—The dye-bath is made with 2 to 10 per cent. of acetic acid, and the colouring matter. The dyeing is done at from 120deg. to 150deg. F. Applicable to the Eosines, Erythrosine, Phloxine, Rhodamines, Croceines. Azo Eosine, Violamines, Oranges, Yellows, Bordeaux, Ponceaus, Scarlets.

METHOD D—The dye-bath is made with 5 to 10 per cent. phosphate of soda, 1 to 2½ per cent. of soap, and the colouring matter. The dyeing is done at 212deg. F. Applicable to all Benzidine, etc., colours named in Cotton Dyeing, Methods E and

F, and in Wool Dyeing, Method E. This method is also applicable to half silk goods.

METHOD E—The dye-bath is made with 10 per cent. of common salt, a slight trace of acetic acid, and the colouring matter. The dyeing is done at from 180deg. to 212deg. F. Applicable to Chrysophenino, Titan Pinks, Reds and Yellows, Jet Black, Benzo-Browns, Hessian Yellow, etc.

METHOD F—Same process and colours as in Method C, Wool Dyeing.

METHOD G—The silk is steeped in a bath of basic sulphate of alumina all night, then dyed the next day in the colouring matter. Applicable to the Alizarine colours and similar colouring matters.

JUTE DYEING.

Essentially the same as Cotton Dyeing.

CALICO PRINTING.

METHOD A—Make the printing colour with 10lb. starch, 3½ gallons water, 1½ gallons acetic acid 8deg. Tw., 4lb. tannic acid; boil up, allow to cool, and then add 36oz. colouring matter dissolved in 1½ gallons of water and ¼ gallon of acetic acid. Print, steam at 14lb. pressure for one hour, run through a bath of tartar emetic, wash well, and dry. The proportion of colouring matter may be varied according to shade. Applicable to all the basic colouring matters called tannic colours by the printer, named in Cotton Dyeing, Method A.

METHOD B—Prepare a printing colour with 10 gallons of water, 2lb. olive oil, 12lb. starch, boil, cool, add 3 quarts acetate of alumina 16deg. Tw., 10oz. tartaric acid, 5 quarts gum tragacanth liquor, and the colouring matter in quantity according to shade required. Generally applicable to all azo-colouring matters and to the Alizarine colours (which see). Instead of acetate of alumina, acetate of chrome or iron may be used.

METHOD C—Make a printing colour with colouring matter 10oz., water one gallon, starch 2lb., gum tragacanth liquor one quart, soap 10oz.; print, steam for one hour, and wash. Applicable to the Benzidine, etc., colours named in Wool Dyeing, Method E. The results cannot be regarded as wholly satisfactory.

It is impossible to give many standard methods of applying coal tar colours in calico printing in this place.

WOOL AND SILK PRINTING.

Fabrics with the two fibres either separately or mixed together can be printed with printing colours made in the same way.

METHOD A—Prepare the printing colour with ½oz. of colour, 1½ gallons water, 11oz. starch, and 5½oz. glue. Boil up, print, and steam for half an hour without pressure. Applicable to all tannic colours named in Method A, Cotton Dyeing.

METHOD B—Prepare the wool or silk cloth for printing by mordanting strongly in a bath of alum, using any of the ordinary methods. (See Method F, Wool Dyeing, and Method G, Silk Dyeing, above.) Print on a colour made with 2 to 5oz. colour, 1½ gallons water, 10½oz. starch, 5½oz. glue, 2oz. tartaric acid. Print and steam for half an hour without pressure. Applicable to all acid and azo colours named in Methods B and D, Wool Dyeing.

METHOD C—Prepare the wool by passing through a weak bath of chloride of lime, followed by one of hydrochloric acid; after washing and drying the cloth is ready for printing. Silk can be bleached by any of the usual methods. Make the printing colour with 2oz. gum tragacanth, 10oz. starch made into a thickening in the usual way with 3½ pints of water, 6oz. sulphate of alumina, 3oz. oxalic acid, 1 pint water, ½ to 2lb. colour. Print, steam for one hour at 7½lb. pressure, soap lightly once or twice, wash, and dry. Instead of 6oz. sulphate of alumina, a similar quantity of copperas, or 1lb. acetate of chrome 32deg. Tw., may be used. Applicable to all the colouring matters named in Method F, Wool Dyeing.

These descriptions of the methods of applying the coal tar colours, while being based on practical experience, should be regarded more in the nature of hints than as absolute recipes.

Mikado Brown—

(LEONHARDT AND CO.)—Similar in composition to Mikado Orange. (1888.)—*English Patent* 2,664, 1888.—Made in various brands, G, B, and M.

Mikado Brown G—

Is a *purple-brown powder*, slightly soluble in cold water, easily in hot water to a deep yellow-brown solution, very slightly soluble in alcohol, slightly soluble in acetic acid, soluble in sulphuric acid to a deep violet solution, from which on diluting with water a dark brown precipitate falls down. Hydrochloric acid added to the aqueous solution throws down an olive-brown precipitate; caustic soda changes the colour of the

aqueous solution to an orange-brown.—
Dyes cotton from a boiling salt bath, silk
from a faintly acid salt bath orange-brown
shades; acids turn them dark brown,
alkalies redden. Fast to soap, moderately
fast to light.

Mikado Brown M—

Is a *dark purple - brown powder*,
soluble in water to dark reddish
brown solution, in alcohol to a purple-
brown solution, in acetic acid to a
yellowish brown solution, in strong sulphuric acid to a violet solution, from
which on diluting with water a blackish
brown precipitate falls down. Hydrochloric acid added to the aqueous solution
throws down a blackish brown precipitate;
caustic soda turns the colour redder.—
Dyes cotton from a boiling salt bath,
silk from a faintly acid salt bath deep seal-
browns. Fast to soaping, moderately
fast to light, darkened by acids, and
reddened by alkalies.

Mikado Orange—

(LEONHARDT AND CO.)—

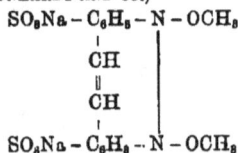

$$SO_3Na - C_6H_3 - N - OCH_3$$
$$|$$
$$CH$$
$$\|$$
$$CH$$
$$|$$
$$SO_3Na - C_6H_3 - N - OCH_3$$

This formula is open to doubt; the
various brands G, R, and RR of Mikado
Orange are obtained by varying the alkyl
groups. (1888.) *English Patent* 2,664,
1888.

Mikado Orange G —

Is a *dark orange-red powder*, soluble
in water to an orange solution, in alcohol
to an orange solution, in acetic acid to a
brownish orange turbid solution, in strong
sulphuric acid to a deep purple solution;
on diluting with water turns amber
coloured. Hydrochloric acid added to the
aqueous solution throws down a dark
brown precipitate; caustic soda throws
down an orange precipitate from the
aqueous solution.—*Dyes* cotton from a
boiling salt bath, and silk from a faintly
acid bath a fine orange. Moderately fast
to light, fast to soaping, reddened by
alkalies, turned brown by acids.

Mikado Orange RR—

Is a *dark red powder*, soluble in water
to an orange solution, in alcohol to a dark
orange solution, in acetic acid to a dull
brownish orange solution, in strong sulphuric acid to a dark blue solution, from

which on diluting with water a blackish
brown precipitate falls down. Hydrochloric acid added to the aqueous solution throws down a blackish brown
precipitate; caustic soda turns the colour
of the aqueous solution slightly redder.—
Dyes cotton from a boiling salt bath, silk
from a faintly acid bath reddish orange.
Moderately fast to light, fast to soaping,
turned brown by acids, reddened by
alkalies.

Milling Red B—

(CASSELLA.)—*Crimson-red powder*, soluble in water to a crimson-red solution,
in alcohol and acetic acid to crimson-red
solutions, in strong sulphuric acid to a
fine violet solution, from which on diluting with water a crimson-red precipitate
rapidly collects on the top of a scarlet
solution. Hydrochloric acid added to the
aqueous solution throws down a fine
crimson-red precipitate; caustic soda
changes the colour of the aqueous solution
to a scarlet.—*Dyes* wool and silk from
acid baths a fine deep red. Fast to acids,
alkalies, and soap; moderately fast to
light.

Milling Red R—

(CASSELLA.) — *Dark purple - brown
powder*, soluble in water to a crimson
solution, in alcohol it is slightly soluble,
in acetic acid it is readily soluble to a
deep crimson solution, in strong sulphuric
acid to a bright blue solution, from which
on adding water a deep crimson precipitate falls out. Hydrochloric acid added
to the aqueous solution throws down a
crimson precipitate; caustic soda changes
the colour to a scarlet.—*Dyes* wool and
silk from acid baths deep wine-red. Fast
to acids, alkalies, and soap; moderately
fast to light.

Milling Yellow O—

(CASSELLA AND CO.)—(1890).—*Orange-
yellow powder*, soluble in water to an
amber-coloured solution, in strong sulphuric acid to a crimson solution; on
diluting with water a brownish yellow
precipitate is obtained. Hydrochloric
acid added to the aqueous solution gives
a brownish yellow gelatinous precipitate;
caustic soda gives a yellow precipitate.—
Dyes wool from an acid bath bright
orange-yellow. Fast to dilute acids, alkalies, and light; not quite fast to soaping.

Mimosa—

(GEIGY.)—Prepared by the action of
ammonia on diazotised polychromine.
(1890.) *English Patent* 1,771, 1890.—
Brownish - yellow powder, soluble in

water to a yellowish orange solution, in alcohol to a pale yellow solution, in acetic acid to a cloudy orange solution ; in strong sulphuric acid to a brownish yellow solution, from which on diluting with water an orange precipitate falls down. Hydrochloric acid added to the aqueous solution throws down a flocculent orange precipitate; caustic soda turns the colour of the aqueous solution to a scarlet-orange. —*Dyes* wool and cotton from a boiling salt bath bright yellows, turned red by acids, brownish orange by alkalies. Not quite fast to soap on wool, fast on cotton; not fast to light on cotton, rather faster on wool.

Muscarin—

(DURAND AND HUGUENIN.)—Chloride of dimethyl - phenyl - *p*-ammonium - *B* - oxy - naphtholazine

$$ClN(CH_3)_2C_6H_3\big\langle{}^N_O\big\rangle C_{10}H_5OH$$

obtained by acting with nitrate of nitroso-dimethylaniline upon *a*-dioxynaphthalene. (1885.) — *Dark brown - violet powder*, slightly soluble in cold, easily in hot water to a blue-violet solution ; boiling the aqueous solution with zinc dust decolourises it ; the colour returns on exposure to air. Soluble in strong sulphuric acid to a blue-green solution ; on diluting with water this turns first blue, then violet, finally a violet precipitate is obtained. Hydrochloric acid added to the aqueous solution gives a blue-violet precipitate. Caustic soda turns the colour of the aqueous solution yellow-brown.—*Dyes* cotton mordanted with tannin and tartar emetic deep shades of indigo blue, fast to acids, light, and washing.

N

Naphthalene —

$C_{10}H_8$

An aromatic hydrocarbon, found very largely in coal tar, and is a most important body for the production of the coal tar colours. It forms *large colourless lustrous crystalline plates*, having a faint but peculiar odour. It melts at 80deg. C. to a colourless liquid, which solidifies into a crystalline mass on cooling; it boils at 218deg. C., and its vapour readily condenses into a lustrous white flaky mass. It is combustible when heated, burning with a luminous but smoky flame; insoluble in cold, but slightly so in hot water, soluble in alcohol, ether, benzol, and other solvents. Naphthalene by the action of nitric acid is converted into nitro-naphthalene $C_{10}H_7NO_2$, dinitronaphthalene $C_{10}H_6(NO_2)_2$; these by reducing agents can be converted into the amido compounds, naphthylamines. The constitution of naphthalene is represented by the formula :

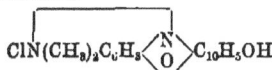

There is certainly a benzene nucleus in naphthalene because it can be made to yield benzene compounds, and other reactions indicate that the molecule of naphthalene is symmetrical in construction, and is represented by the above formula, as well as a formula can show the grouping of atoms in the molecule of a chemical compound. By replacing the hydrogen of naphthalene by radicles and elements numerous substitution derivatives are formed ; the number of possible ones is very great, and many of these are known, and are of great technical value. The monoderivatives exist in two isomeric forms, due to the position of the replacing radicle in the molecule ; the places marked 1, 4, 5, 8 in the above formula are similar to each other, but are different to those marked 2, 3, 6, 7, which are also similar to each other. The derivatives where the radicle is in the positions $1=4$ $=5=8$ are called *a* derivatives, those in which the radicle occupies the position $2=3=6=7$ are called *b* derivatives ; thus there are two naphthols, *a*-naphthol and *b*-naphthol, and two naphthylamines, *a*-naphthylamine and *b*-naphthylamine, as shown in the following formulæ—

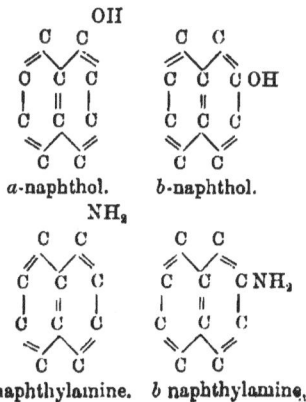

a-naphthol. *b*-naphthol.

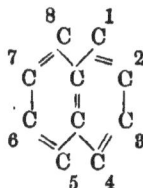

a-naphthylamine. *b* naphthylamine.

The naphthols and naphthylamines are very valuable agents in the production of dyes.

Of the diderivatives of naphthalene there are ten possible isomeric modifications (see DIOXYNAPHTHALENE), some of which are known, others have not as yet been isolated.

Naphthalene Yellow—
Syn. of Naphthol Yellow.

Naphthion Red—
Syn. of Orchil substitute.

Naphthionic Red—

Nitrobenzene - azo - *a* - naphthylamine-*a*-sulphonic acid

$$NO_2C_6H_4N : NC_{10}H_5NH_2HSO_3$$

obtained by diazotising nitraniline and combining with naphthionic acid. *Paste,* soluble in water, precipitated by acids and alkalies. Soluble in strong sulphuric acid to a magenta coloured solution, turning brown, and giving a precipitate on diluting with water.—*Dyes* wool and silk from acid baths an orchil red, fast to acids and washing.

Naphthol Sulphonic Acids—

When the naphthols are treated with sulphuric acid, there is first formed napthyl sulphates $(C_{10}H_7)_2 SO_4$, which on being heated pass into sulphonic acids of naphthol; first the mono, then subsequently, on further action, the di and tri sulphonic acids are formed. Of these bodies several possible isomers are capable of being formed, and many are known. They vary considerably in their technical value as producers of colouring matter. Which isomer is formed depends upon several factors, relative proportion of the naphthol and sulphuric acid used, kind of naphthol, temperature, time of action, and other factors. In making these, therefore, considerable care has to be taken in regulating the temperature, time of sulphonating, and other points, so as to ensure the formation of the desired sulphonic acid. The most important of these acids are:—

1. MONOSULPHONIC ACIDS, $C_{10}H_6OHSO_3$: *a*-naphthol acid N. W., obtained by Neville and Winter from naphthionic acid; *a*-naphthol acid C. of Cleve; *B*-naphthol acid B. (Bayer's acid); *B*-naphthol acid S. (Schaffer's acid); *B*-naphthol acid F., *English Patent* 12,908, 1886; *B*-naphthol *a*-acid, *English Patent* 1,225, 1881.

2. DISULPHONIC ACIDS, $C_{10}H_5OH(HSO_3)_2$: *a*-naphthol diacid Sch., Schollkopf Aniline and Chemical Co.; *a*-naphthol diacid B., *English Patent* 4,625, 1888; *a*-naphthol

diacid L., *English Patent* 11,818, 1887; *a*-naphthol diacid S., *English Patent* 13,665, 1890; *B*-naphthol diacid R.; *B*-naphthol diacid G.—the last acids are exceedingly valuable in the preparation of azo-colours — *English Patent* 1,715, 1878 ; *B*-naphthol-delta-diacid, *English Patent* 8,265, 1887; *B*-naphthol gamma diacid, *English Patent* 816, 1884.

3. TRISULPHONIC ACIDS. *B*-naphthol triacid M., *English Patent* 2,544, 1882.

It has not been considered necessary to give detailed accounts of these different acids. Many patents have been taken out for the production of naphthol acids; but the question arises whether some of these are not identical with one another. The sulpho acids of naphthol require investigation.

Naphthols—

$$C_{10}H_7OH$$

Of this derivative of naphthalene there are two modifications: one known as the *alpha*, the other as *beta* naphthol. By treating naphthalene with strong sulphuric acid it is converted into *a* and *b* monosulphonic acids $C_{10}H_7SO_3H$, and these when fused with caustic soda yield the corresponding naphthols. When the sulphonation is conducted at a low temperature a mixture of both naphthylsulphonic acids is obtained; these can be separated by converting them into barium salts, and fractionally crystallising, the *a* acid salts being much more soluble than the *b* acid salts. When the sulphonation is carried on at from 150deg. to 170deg. C. only the *b* acid is formed.

A - naphthol crystallises in colourless prisms, melting at 94deg. C., boiling at 278deg. to 280deg. C., nearly insoluble in cold water, more soluble in hot water, readily soluble in alcohol and ether.

B-naphthol crystallises in small rhombic plates, melting at 122deg. C., boiling at 285deg. to 290deg. C., sublimes easily, sparingly soluble in boiling water, easily soluble in alcohol and ether.

Both varieties are soluble in solutions of caustic soda, from which they are reprecipitated on the addition of an acid. By treatment with sulphuric acid both naphthols are converted into sulphonic acids—mono, di, and tri—of which several isomeric varieties are known (see NAPHTHOL SULPHONIC ACIDS). By treatment with nitric acid nitro-naphthols are formed which are useful as colouring matters (see NAPHTHOL YELLOW). Both naphthols meet with extensive use in colour making; of the two the *beta*-naphthol is the most valuable, yielding by far the most brilliant colours. The *alpha*-naphthol gives darker and duller shades. The azo-colouring matters into which the naphthols themselves enter are mostly insoluble in water, but those containing their sulphonic acids, especially the disulphonic acids, are readily soluble.

Naphthol Black—

(CASSELLA.)—Sodium salt of naphthalene disulphonic - acid - azo - naphthalene - azo - naphthol disulphonic acid

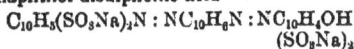

$$C_{10}H_5(SO_3Na)_4N : NC_{10}H_6N : NC_{10}H_4OH$$
$$(SO_3Na)_2$$

obtained by diazotising amido-azo-naphthalene disulphonic acid, and combining with *B*-naphthol disulphonic acid R. (1885.) *English Patent* 9,214,1885.—*Blue-black powder*, soluble in water to a violet solution, in strong sulphuric acid to a green solution; on diluting with water this turns first blue, then a red-violet precipitate is thrown down. Adding hydrochloric acid to the aqueous solution throws down a red-violet precipitate. Caustic soda gives a blue precipitate soluble in much water.—*Dyes* wool and silk in acid baths fine blue-black shades, very fast to light, acids, and washing. By adding a little yellow or naphthol green a jet black can be got. Can be used in wool and silk printing, giving blues and blacks. Also known as Brilliant Black (BADISCHE).

Naphthol Black 3B—

(CASSELLA.) — (1890). — *Bronzy crystalline powder*, soluble in water to a dark blue solution, insoluble in alcohol, soluble in strong sulphuric acid to a greenish black solution; on diluting with water it turns plum coloured. Hydrochloric acid added to the aqueous solution causes little change ; caustic soda turns it a little bluer.—*Dyes* wool and silk from acid baths deep navy blue, fast to acids, light, alkalies, moderately so to soaping.

Naphthol Green B—

(CASSELLA AND CO.).—Ferrous sodium salt of nitroso-*B*-naphthol monosulphonic acid

$$C_{20}H_{10}N_2O_{10}S_3FeNa_2$$

(1888.) — *English Patent*, made by the action of nitrous acid on *B*-naphthol monosulphonic acid S, and combination with ferrous sodium salts.—*Dark olive-green powder*, soluble in water to a yellow-green solution, in strong sulphuric acid to a yellow-brown solution. Addition of hydrochloric acid to the aqueous solution produces no change; caustic soda turns

the colour to a blue-green.—*Dyes* wool and silk in an acid bath olive-green shades, very fast to light, air, soaping, and acids.

a-Naphthol Orange—
Syn. of Orange I.

b-Naphthol Orange—
Syn. of Orange II.

Naphthol Violet—

Is a mixture of several compounds. Obtained by treating nitrosodimethylaniline with *B* - naphthol. (1881.)—MELDOLA. *Chem. Soc. Journ.*, 1881, p. 57.

Naphthol Yellow—

(ROBERTS, DALE AND CO.)—Alkali salt of dinitro-*alpha*-naphthol

$$C_{10}H_5(NO_2)_2ONa + H_2O$$

obtained by treating *alpha*-naphthol sulphonic acid with nitric acid. (1864.) MARTIUS.—*Yellow needles*, not easily soluble in water; hydrochloric acid decolourises the aqueous solution, precipitating the free acid; caustic soda gives orange-coloured precipitates.—*Dyes* wool and silk golden yellows, which are not fast to washing or acids; easily volatile, so that heating a little of the dyed fibre causes the colour to become discharged. Known also as Manchester Yellow, Martius Yellow, Naphthalene Yellow, Naphthylamine Yellow, Gold Yellow.

Naphthol Yellow S—

Soda salt of dinitro-*a*-naphthol-sulphonic acid

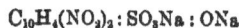

$$C_{10}H_4(NO_2)_2 : SO_3Na : ONa$$

obtained by treating *a*-naphthol trisulphonic acid with nitric acid. (1879.) CARO. — *Orange - yellow powder*, easily soluble in water to a yellow solution, acids turn the colour paler, alkalies give in dilute solutions flocculent precipitates.—*Dyes* wool and silk fine gold yellow shades in acid bath. Known also as Acid Yellow, Acid Yellow S.

Naphthorubin—

(BAYER.)—Sodium salt of *a*-naphthylamine-azo-disulphonic acid

$$C_{10}H_7N : NC_{10}H_4OH(SO_3Na)_2$$

obtained by diazotising naphthylamine, and combining with *a*-naphthol-disulphonic acid. GURCKE. (1886.)—*English Patent* 15,716, 1885.—*Brown powder*, soluble in water to a crimson solution, and in strong sulphuric acid to a violet solution; on diluting with water it turns red. Hydrochloric acid does not change the colour of the aqueous solution, while caustic soda turns it a yellow-red.—*Dyes* wool and silk in acid bath dark shades of bluish red.

Naphthylamines or Amidonaphthalenes
$$C_{10}H_7NH_2$$

There are two possible bodies having this formula, and both are known and are of considerable importance in colour manufacturing. The *alpha* naphthylamine is made on a large scale by the reduction of *a*-nitro-naphthalene with iron filings and hydrochloric acid. It is obtained in *colourless crystals*, having a peculiar disagreeable odour, melting at 50deg. C., and vaporising at 300deg. C. It is insoluble in water, soluble in acids forming the naphthylamine salts, which are well characterised, are easily soluble in water; on addition of oxidising agents, such as ferric chloride or chromic acid, to the aqueous solutions, a purple-red powder of naphthamein,

$$C_{10}H_9NO$$

is obtained. *Alpha*-naphthylamine is converted by sulphuric acid first into the sulphate and afterwards into the sulphonic acids.

The *beta*-naphthylamine is obtained on the large scale by heating in a closed vessel a mixture of *beta*-naphthol, caustic soda, and ammonium chloride. The reaction lies essentially between the naphthol and the ammonia liberated from the other constituents, and is expressed by the equation

$$C_{10}H_7OH + NH_3 = C_{10}H_7NH_2 + H_2O$$
$$\text{B-naphthol.} \qquad \text{B-naphthylamine.}$$

It forms *pearly crystalline masses, colourless when pure ;* the ordinary commercial product usually has a pinkish colour, free from smell, melting at 112deg. C. Insoluble in water, soluble in acids to form the salts. With sulphuric acid it forms several sulphonic acids.

Both naphthylamines are used in the manufacture of colours, the *beta* variety giving the brightest colours.

Naphthylamine Sulphonic Acids—

There are several isomeric sulphonic acids of the two—*a* and *b*—naphthylamines. These are obtained in several ways: by sulphonation of the amine, or from the corresponding naphthol acid by the action of ammonia. The different isomeric acids have a different value in the production of colouring matters; some are of great technical value, others are not. The entire possible acids have not yet been completely isolated, and much work yet remains to be done before the various acids now known can be

properly differentiated one from the other. This is rendered more complicated by the fact that in the process of manufacture pure acids are seldom obtained ; usually more or less complex mixtures of several sulphonic acids, of which, perhaps, one will predominate over the rest. The acids are distinguished one from the other by their physical appearance, solubility in various solvents, solubility of the alkali and alkaline earth salts ; and usually they are separated one from the other by taking advantage of varying solubilities of their salts.

The following acids are well known and largely used in the production of dyestuffs :—

1. Monosulphonic acids : *A*-naphthylamine *a* acid (Piria's Naphthionic acid), made by acting with sulphide of ammonia on *a*-nitronaphthalene by action of fuming sulphuric acid on *a*-naphthylamine. It has the formula $C_{10}H_6aNH_2\,HSO_3$. *A*-naphthylamine acid (LAURENT'S) obtained by the reduction of *a*-nitronaphthalene monosulphonic acid, $C_{10}H_6aNH_2\,HSO_3$.

B-naphthylamine acid Br., $C_{10}H_6BNH_2$ HSO_3.

B-naphthylamine acid D. (Dahl's acid), *English Patents* Nos. 7,712, 7,713, 1884. $C_{10}H_6BNH_2\,HSO_3$.

B-naphthylamine acid *delta* (BAYER'S).

Napthylamine Black D—

(CASSELLA.)—(1890.) *British Patent.*—*Blackish brown powder*, soluble in water to a dark purple solution, in alcohol and acetic acid to a dark violet solution, in strong sulphuric acid to a navy blue solution; on diluting with water a reddish purple precipitate is obtained. Hydrochloric acid throws down a reddish purple precipitate from the aqueous solution ; caustic soda gives a dark bluish black precipitate.—*Dyes* wool and silk from a neutral bath violet-red to reddish black shades ; fast to acids, light, and alkalies ; not fast to soaping.

Naphthylamine Yellow—
Syn. of Naphthol Yellow.

Naphthylene Blue R Crystals—
Syn. of Meldola's Blue.

Naphthylene Red—
(BADISCHE.)—

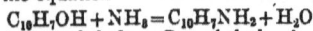
$$\text{Tetrazonaphthyl} \Big\langle \begin{array}{l} \text{sodium naphthionate,} \\ \text{sodium naphthionate.} \end{array}$$

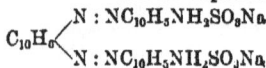
$$C_{10}H_6 \Big\langle \begin{array}{l} N : NC_{10}H_5NH_2SO_3Na \\ N : NC_{10}H_5NH_2SO_3Na \end{array}$$

obtained by azotising naphthalene diamine and combining with sodium naphthionate. (1888.)—*English Patent.*— *Dyes* cotton from an alkaline soap bath scarlet, fast to washing and soap, not to light, turned blue by acids.

Narcein—

(DURAND AND HUGUENIN.)—Sodium salt of benzene sulphonic acid hydrazo-naphthol sulphonic acid

$$C_6H_4SO_3NaNH—NSO_8NaC_{10}H_6OH$$

made by treating Orange II. with sodium bisulphite.(1879.)—PRUDHOMME.—*Orange-red powder*, easily soluble in water to a yellow solution, in strong sulphuric acid to a yellow solution ; on diluting with water and warming, sulphur dioxide is evolved. No change is produced on adding hydrochloric acid to the aqueous solution, but caustic soda turns it brown-red. Used chiefly for calico printing.

Navy Blue—

Syn. of Water Blue.

Neutral Blue—

(CASSELLA.)—Phenyldimethyl-*p*-amidopheno-naphthazonium chloride

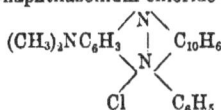

$$(CH_3)_2NC_6H_3 \underset{\underset{\diagdown}{N}}{\overset{\overset{\diagup}{N}}{}} \begin{array}{c} C_{10}H_6 \\ \\ C_6H_5 \end{array} Cl$$

acting with nitrate of nitroso-dimethylaniline upon phenylnaphthylamine. (1882.) —O. N. WITT.—*Brown powder*, easily soluble in water to a violet solution, in alcohol to a red-violet solution, in strong sulphuric acid to a brown-violet solution ; on diluting with water this turns violet. Hydrochloric acid turns the colour of the aqueous solution bluer ; caustic soda gives a violet precipitate.—*Dyes* cotton mordanted with tannin and tartar emetic, wool and silk in neutral baths, red shades of blue.

Neutral Red—

(CASSELLA.)—Hydrochlorate of dimethyl-diamidotoluphenazine

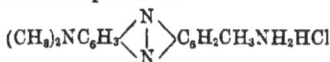

$$(CH_3)_2NC_6H_3 \underset{\underset{\diagup}{N}}{\overset{\overset{\diagdown}{N}}{}} C_6H_2CH_3NH_2HCl$$

obtained by acting with nitrate of nitrosodimethylaniline upon *m*-toluylenediamine. (1879.)—O. N. WITT.—*Dark greenish black powder*, easily soluble in water to a crimson solution, in alcohol to a magenta-red solution, having a weak brown-red fluorescence ; soluble in strong sulphuric acid to a green solution ; on diluting with water this turns first blue, then magenta-

red. Hydrochloric acid turns the colour of the aqueous solution a pure blue ; caustic soda gives a yellow-brown precipitate.—- *Dyes* cotton mordanted with tannin and tartar emetic, wool and silk in neutral baths, bluish reds, moderately fast to light, acids and washing. Known also as Toluylene Red.

Neutral Violet—

(CASSELLA AND CO.)—Hydrochlorate of dimethyl-diamidophenazine

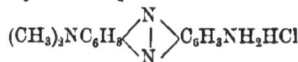

$$(CH_3)_2NC_6H_3 \underset{\underset{\diagup}{N}}{\overset{\overset{\diagdown}{N}}{}} C_6H_3NH_2HCl$$

obtained by acting with nitrate of nitro-dimethylaniline upon *m*-phenylenediamine. (1879.)—O. N. WITT.—*Greenish black powder*, easily soluble in water to a violet-red solution, in strong sulphuric acid to a green solution ; on diluting with water it turns first blue, then violet. Dilute hydrochloric acid added to the aqueous solution has little action ; strong hydrochloric acid turns it blue ; caustic soda gives a brown precipitate.—*Dyes* cotton mordanted with tannin and tartar emetic, wool and silk in neutral baths, red-violet, moderately fast to light, acids and soaping.

New Blue—

Syn. of Meldola's Blue.

Is made in several brands. The New Blue R when pure is dimethyl-amido-naphthophenoxazine chloride, and may be considered the base from which the other brands are made. Very useful for dyeing fast dark blues on cotton ; and can be used in calico printing.

New Coccin—

Syn. of Brilliant Ponceau (CASSELLA).

New Coccin R—

(ACTIENGESELLSCHAFT.)—Syn. of Crystal Scarlet 6R (CASSELLA).

New Grey—

(BAYER.)—A derivative of nitrosodimethylaniline. (1889.)—*Black powder*, slightly soluble in cold, easily in boiling water to a reddish black solution, in strong sulphuric acid to a brownish solution ; on diluting with water a red precipitate is obtained. Hydrochloric acid added to the aqueous solution gives a red precipitate ; caustic soda has no action.—*Dyes* cotton mordanted with tannin and tartar emetic fine reddish greys, fairly fast to light, acids and washing. Can be used in calico printing.

New Green—
(BAYER.)—Syn. of Malachite Green.

New Red L—
(KALLE.)—Syn. of Biebrich Scarlet.

New Red 5R—
Is isomeric with Croceine Scarlet 3B.

New Solid Green 3B—
(SOCY. CHEM. IND., DASLE).—Syn. of
Victoria Green 3B.

New Victoria Green—
Syn. of Brilliant Green.

New Yellow—
(DAYER.)—Syn. of Acid Yellow.

New Yellow—
(BEYER AND KEGEL)—Syn. of Flavaurin.

New Yellow L—
(KALLE.)—Syn. of Acid Yellow.

New Yellow—
(TILLMANNS E. TER. MEER AND CO.)—
Syn. of Curcumein (ACTIENGESELLSCHAFT).

Nicholson Blue—
(See ALKALI BLUE.)

Night Blue—
Syn. of Spirit Blue.

Night Blue—
(BADISCHE. SOCY. CHEM. IND., BASLE.)
—Hydrochlorate of tolyltetramethyltri-
amido-*a*-naphthyldiphenylcarbinol

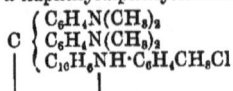

$$C \begin{cases} C_6H_4N(CH_3)_2 \\ C_6H_4N(CH_3)_2 \\ C_{10}H_6NH \cdot C_6H_4CH_3Cl \end{cases}$$

obtained by acting with *p*-tolyl-*a*-naph-
thylamine on tetramethyldiamido-benzo-
phenonechloride. (1883.)—CARO AND KERN.
English Patent 5,008, 1884; 11,159, 1884.
—*Violet bronzy lustrous powder*, soluble
in water with a blue-violet colour, in strong
sulphuric acid with a yellow-brown colour;
on diluting this solution with water it turns
first green, then blue; easily soluble in
alcohol. Hydrochloric acid turns the
aqueous solution first green, then yellow-
brown; caustic soda gives a red-brown
precipitate.—*Dyes* wool from an acid
bath; silk from a broken soap or from an
acid bath; cotton either direct with a little
acetic acid or mordanted with tannin and
tartar emetic. The shades given are fine
greenish blue, retaining their colour in
gaslight, moderately fast to light, air and
washing.

Night Green—
Syn. of Iodine Green.

Nigramine—
(WM. NOETZEL AND CO.) (1890).—*Black
crystalline powder*, slightly soluble in
cold, easily in hot water to a violet-blue
solution; soluble in alcohol to a bright
blue, and in acetic acid to a dark blue
solution. Strong sulphuric acid gives a
brownish black solution, which, on diluting with water, turns a purple-brown.
Hydrochloric acid darkens the colour of
the aqueous solution; caustic soda decolorises and throws down a white precipitate of the base.—*Dyes* tannin mordanted cotton a fine bluish grey, turned
green by acids, brownish red by caustic
soda. Fast to soaping and light. May be
used in calico printing with a tannic
mordant.

Nigrisine—
(POIRRIER.)—(1890.)—EHRMANN. *English Patent* 5,032.—*Black powder*, soluble
in water to a reddish grey solution, in
strong sulphuric acid to a grey solution; on
diluting with water it turns first reddish
grey, then blue-grey. Very slightly
soluble in alcohol. Acids added to the
aqueous solution turn it blue grey.
Alkalies precipitate the nigrisine base as a
black powder, insoluble in water, soluble
in acids.—*Dyes* cotton mordanted with
tannin and tartar emetic fine blue-grey
shades, or it can be dyed by simply boiling in a bath of the colouring matter,
drying and steaming. A bath of bichromate of potash also fixes the colour.
In calico printing it is used with a
tannin-acetic-tartaric acid thickening. It
can be printed on silk and wool in the
same way. The shades are fast to light,
acids and soaping.
Also known as MALTA Grey.

Nigrosine Spirit Soluble—
Syn. of Induline Spirit Soluble.

Nigrosine Water Soluble--
Syn. of Induline Water Soluble.

Nile Blue—
(BADISCHE.)—Sulphate of dimethyl-
phenyl-*p*-ammonium-*a*-amido-naphthoxa-
zine

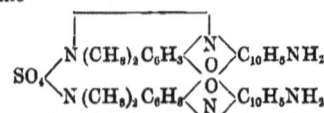

$$SO_4 \begin{cases} N(CH_3)_2C_6H_3\underset{O}{\overset{N}{<}}C_{10}H_5NH_2 \\ N(CH_3)_2C_6H_3\underset{N}{\overset{O}{<}}C_{10}H_5NH_2 \end{cases}$$

obtained by acting with nitrate of nitro-
sodimethyl-*m*-amido-phenol upon *a*-naph-

thylamine. (1888.) — *Greenish bronzy lustrous crystalline powder*, slightly soluble in cold, easily in warm, water to a blue solution, soluble in alcohol to a blue solution, in strong sulphuric acid to a yellow solution ; on diluting with water the colour changes through green to blue. On adding hydrochloric acid to the aqueous solution the hydrochlorate precipitates out in the form of needles, which by transmitted light are violet, by reflected light green ; caustic soda gives a red precipitate, soluble in ether to a brown-orange solution with a dark green fluorescence.— *Dyes* wool and silk in neutral baths ; cotton requires a mordant of tannin and tartar emetic ; fine bright shades of blue are obtained which are moderately fast to light, acids, and washing.

Nitrazine Yellow—

(K. OEHLER.)—Obtained by condensing nitro-xylyl-hydrazine sulphuric acid with dioxy-tartaric acid. (1891.) — *English Patent* No. 10,599, 1890.—*Pale orange-coloured powder*, readily soluble in water to a greenish yellow solution ; in alcohol and acetic acid it gives greenish yellow solutions. Strong sulphuric acid gives an amber solution, which on diluting with water turns greenish yellow. Hydrochloric acid has no action on the aqueous solution, as also has caustic soda.—*Dyes* wool and silk fine bright yellows, fast to acids ; slightly darkened by caustic soda, not fast to soaping, fast to light.

Nitro-colouring Matters—

When organic bodies are treated with nitric acid they frequently undergo nitration ; they have one or other of their elemental constituents replaced by the group NO_2, known as nitroxyl. Some of the nitro-bodies, such as trinitrophenol (Picric acid), $C_6H_2(NO_2)_3OH$; dinitronaphthol (Manchester Yellow), $C_{10}H_5(NO_2)_2OH$; dinitro-naphthol sulphonic acid (Acid Yellow), $C_{10}H_4HSO_3(NO_2)OH$; tetranitro-naphthol (Heliochrysin), $C_{10}H_3(NO_2)_4OH$; hexanitrodiphenylamine (Aurantia), $N_2C_6H_2(NO_2)_3H$, form useful dyestuffs, dyeing yellows or oranges, not remarkable for great fastness to light, air, etc.; but they have great colouring power, and give brilliant shades on wool and silk, for which fibres they are alone applicable. They are dyed in an acid bath with sulphuric acid and Glauber's salt.

Nitroso-Colouring Matters—

When phenols, etc., are acted upon by nitrous acid they take up the group NO (nitrosyl) to form nitrosocompounds ; some

of these form colouring matters of great value. Thus there is dinitrosoresorcin (Resorcin Green), $C_6H_2(NO)_2(OH)_2$; nitroso-*a*-naphthol (Gambine R), $C_{10}H_6(NO)OH$; nitroso-*b*-naphthol(GambineY), $C_{10}H_6NOOH$; nitroso-dioxynaphthalene (Gambine B), $C_{10}H_5NO(OH)_2$; while an isomer of this latter body forms the colouring matter Dioxine. Ferrous-sodium nitroso-*b*-naphthol monosulphonate, Fe $(C_{10}H_4NOOHSO_3Na)_2$, is Naphthol green. With the exception of the last, which dyes wool and silk very fast shades of olive-green from a bath containing sulphuric acid and Glauber's salt, the nitroso-bodies are adjective dyestuffs. Generally they are insoluble, or nearly so, in water, and hence are sent out commercially in the form of pastes ; some of them are slightly soluble in acetic acid, sulphuric acid, and other solvents. They require the fibres to be mordanted, and they give different colours with different mordants ; chrome and iron are the two most useful mordants; alumina does not give fast colours. With chrome browns are usually got ; Gambine R and Y and Dioxine give reddish or yellowish browns, Gambine B gives dark olive-browns, almost approaching a black ; with copperas, Resorcin green, Gambines R and Y, and Dioxine give dark greens ; Gambine B gives very dark olive-green shades. They are all fast to light and resist acids and washing very well.

O

Oenanthinin—

(DURAND AND HUGUENIN.)—Syn. of Amaranth.

Opal Blue—

(SIMPSON, MAULE AND NICHOLSON.)— Hydrochlorate of triphenyl-rosaniline

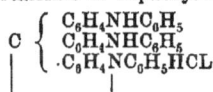

$$C \begin{cases} C_6H_4NHC_6H_5 \\ C_6H_4NHC_6H_5 \\ C_6H_4NC_6H_5HCL \end{cases}$$

obtained by acting with aniline upon rosaniline in the presence of acetic acid. (1860.)—GIRARD AND DE LAIRE, E. C. NICHOLSON. *English Patent*, 24th June, 1862.—*Grey-green powder*, insoluble in water, soluble in alcohol to a blue solution, soluble in strong sulphuric acid to a brownish yellow solution ; on diluting with water a blue precipitate is obtained. Hydrochloric acid added to the aqueous solution has little action ; caustic soda turns it brownish red.—*Dyes* cotton from a soap and alumina mordant, wool and silk direct fine greenish blues ; very fast to light and acids.

Opal blues vary much in composition; some makes are the sulphate and acetate of triphenyl-rosaniline; others contain triphenyl-para-rosaniline. The various shades are produced by varying degrees of phenylation of the rosaniline base. Known also as Aniline Blue Spirit Soluble, Gentiana Blue 6B, Spirit Blue O, Night Blue.

Opal Blue—
Syn. of Spirit Blue.

Opal Blue—
Syn. of Water Blue.

Orange I.—

(WILLIAMS, THOMAS AND DOWER.)— Sodium salt of sulphanilic acid-azo-naphthol

$$C_6H_4SO_3NaN : NC_{10}H_6OH$$

obtained by diazotising sulphanilic acid and combining with a-naphthol. (1876.) WITT. *Chem. Soc. Journ.*, 1879, p. 184. —*Red-brown powder*, soluble in water to an orange-red solution, and in strong sulphuric acid to a magenta-red solution; on diluting with water this turns red-brown. Hydrochloric acid gives a brownish precipitate from the aqueous solution; caustic soda turns the colour of this solution scarlet.—*Dyes* wool and silk in acid baths a fine orange; moderately fast to light and washing, but sensitive to acids and alkalies. Also known as Orange I. (POIRRIER), a-Naphthol Orange, Tropæolin OOO No. 1 (original name).

Orange II.—

(WILLIAMS, THOMAS AND DOWER.)— Sodium salt of sulphanilic acid—azo-B-naphthol

$$C_6H_4SO_3NaN : NC_{.6}H_6OH$$

made by diazotising sulphanilic acid and combining with B-naphthol. (1876.) WITT. *Chem. Soc. Journ.*, 1879, p. 184. —*Yellowish red powder*, soluble in water with a reddish yellow colour, in strong sulphuric acid to a crimson solution; on diluting with water a brownish yellow precipitate is obtained. The same precipitate is formed on adding hydrochloric acid to the aqueous solution; caustic soda turns the colour of the aqueous solution brown.—*Dyes* wool and silk in acid baths bright oranges. Fairly fast to light, air, and washing; but sensitive to acids and alkalies. Also known as Orange No. 2 (POIRRIER), B-Naphthol Orange, Tropæolin OOO No. 2 (original name), Mandarin, Mandarin G Extra (ACTIEN-GESELLSCHAFT), Chrysaurein, Gold Orange (BAYER).

Orange III., or Orange No. 3—

(POIRRIER.)—Sodium salt of m-nitro-benzene-azo-B-naphthol disulphonic acid R

$$C_6H_4(NO_2)N : NC_{10}H_4(OH)(SO_3Na)_2$$

obtained by treating diazo-m-nitrobenzene with B-naphthol disulphonic acid R. (1878.)—ROUSSIN AND POIRRIER. STEBBINS. *Chem. News*, vol. 43, p. 58.— *Red-brown powder*, soluble in water to a reddish yellow solution, in strong sulphuric acid to an orange-yellow solution; on diluting with water an orange-yellow precipitate first forms, this on further addition of water dissolves to a yellow solution. Addition of hydrochloric acid to the aqueous solution gives an orange-yellow precipitate; soluble in much water to a yellow solution; caustic soda turns the colour of the solution brownish. —*Dyes* wool or silk in acid baths orange shades, only moderately fast to light and acids.

Orange III.—

(POIRRIER.)—Sodium salt of sulphanilic acid azo-dimethylaniline

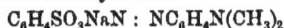

$$C_6H_4SO_3NaN : NC_6H_4N(CH_3)_2$$

made by diazotising sulphanilic acid and combining with dimethylaniline.(1875.)— *Orange-yellow powder*, soluble in water with an orange colour, in strong sulphuric acid to a brown solution; on diluting with water a pink solution is obtained. Addition of hydrochloric acid turns the aqueous solution red; the orange is very sensitive to acids, and hence forms one of the best indicators for mineral acids in volumetric analysis; caustic soda turns the colour of the aqueous solution yellow.—*Dyes* wool and silk in weak acid baths an orange. Not fast to acids, which turn the dyed fibres red, or alkalies, which turn them yellow. Also known as Orange No. 3 (POIRRIER), Methyl Orange, Dimethyl Orange, Dimethylamine Orange, Tropæoline D, Helianthin (BADISCHE).

Orange IV.—

(BADISCHE.)—Syn. of Acid Yellow.

Orange G—

(ACTIENGESELLSCHAFT AND MEISTER, LUCIUS AND BRUNING.) — Sodium salt of benzene-azo-B-naphthol disulphonic acid

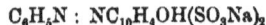

$$C_6H_5N : NC_{10}H_4OH(SO_3Na)_2$$

obtained from diazo-benzene and B-naphthol disulphonic acid G. (1878.)— *Yellowish red powder*, soluble in water to an orange-yellow solution, in strong sulphuric acid to a cherry-red solution;

on diluting with water a reddish yellow solution is obtained. Hydrochloric acid does not change the colour of the aqueous solution ; caustic soda produces an orange solution.—*Dyes* wool and silk in acid baths an orange-yellow, and giving moderately fast shades.

Orange GS—

Syn. of Acid Yellow (1).

Orange GT—

(BAYER.)—Sodium salt of toluene-azo-*B*-naphthol monosulphonic acid

$$C_6H_4CH_3N : NC_{10}H_6OHSO_3Na$$

obtained by treating diazotoluene with *B*-naphthol monosulphonic acid S. (1879.) —*Scarlet-red powder,* soluble in water to an orange-yellow solution, in strong sulphuric acid to a magenta-red solution ; on diluting with water a brown precipitate forms. The same is obtained on adding hydrochloric acid to the aqueous solution ; caustic soda turns the colour of the aqueous solution dark brown.—*Dyes* wool and silk in acid baths an orange-yellow. Fairly fast to light and soaping.

Orange MN—

(SOCY. OF CHEM. IND., BASLE.)—Syn. of Metanil Yellow.

Orange M—

(SOCY. OF CHEM. IND., BASLE.)—Syn. of Acid Yellow (1).

Orange N—

(POIRRIER.)—Sodium salt of p-toluene-o-sulphonic acid-azo-diphenylamine

$$C_6H_3SO_3Na.CH_3.N : N.C_6H_4NH.C_6H_5$$

(1879.)—ROUSSIN AND POIRRIER. *English Patent* 4,491, 1878.—*Yellowish red powder,* soluble in water to a yellow solution, in strong sulphuric acid to a blue-green solution ; on diluting with water a steel-blue precipitate falls down. Hydrochloric acid added to the aqueous solution throws down a steel-blue precipitate ; caustic soda.—*Dyes* wool and silk in acid baths a fine orange. Also known as Jaune N, Curcumein.

Orange No. 1—

(POIRRIER.)—Syn. of Orange I.

Orange No. 2—

(POIRRIER.)—Syn. of Orange II.

Orange No. 3—

(POIRRIER.)—Syn. of Orange III.

Orange F̄

(BADISCHE.)—Sodium salt of xylene-sulphonic acid-azo-*B*-naphthol

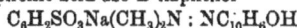

$$C_8H_2SO_3Na(CH_3)_2N : NC_{10}H_6OH$$

obtained by diazotising xylidine mono-sulphonic acid and combining with *B*-naphthol. (1877.)—*Scarlet-red powder,* soluble in water to an orange solution, in strong sulphuric acid to a bluish red solution ; on diluting with water a brown precipitate falls. Addition of hydrochloric acid to the aqueous solution throws down a brown-red precipitate ; caustic soda turns the colour of the aqueous solution a brownish yellow.—*Dyes* wool and silk in acid baths fine orange. Also known as Orange RR (SOCY. CHEM. IND., BASLE).

Orange R—

(SOCY. CHEM. IND., BASLE.)—Syn. of Orange T.

Orange RR—

(SOCY. CHEM. IND., BASLE.)—Syn. of Orange R.

Orange T—

(KALLE.)—Sodium salt of o-toluene sulphonic acid-azo-*B*-naphthol.

$$C_6H_3SO_3NaCH_3N : NC_{10}H_6OH$$

obtained by azotising o-toluidin monosulphonic acid and combining with *B*-naphthol.—*Brick-red powder,* soluble in water to a reddish yellow solution, in strong sulphuric acid to a crimson solution ; on diluting with water a yellow-brown flocculent precipitate is obtained. The same precipitate is obtained by adding hydrochloric acid to the aqueous solution ; caustic soda turns the aqueous solution red-brown.—*Dyes* wool and silk in acid baths orange. Also known as Mandarin GR (ACTIENGESELLSCHAFT), and Orange R (SOCY. CHEM. IND., BASLE).

Orcellin—

(HENRIET, ROMAN AND VIGNON.)— Picramic acid-azo-resorcin

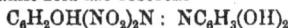

$$C_6H_2OH(NO_2)_2N : NC_6H_3(OH)_2$$

obtained by diazotising picramic acid and combining with resorcin. (1877.)—GRIESS. *English Patent* 3,698, 1877. Not now in use.

Orcellin—

(LEONHARDT.)—(1890.) — *Violet-black powder,* soluble in water to a purple solution, in alcohol to an amber-coloured solution, in strong sulphuric acid to a violet-red solution ; on diluting with water a purplish blue precipitate is obtained. Hydrochloric acid added to the aqueous solution throws down a bluish purple precipitate ; caustic soda has no action.— *Dyes* wool and silk from acid baths dull violet-red shades. Fast to acids, alkalies, and light ; moderately fast to soaping.

Orcellin No. 4—
Syn. of Fast Red A.

Orchil Substitute—
(ACTIENGESELLSCHAFT AND POIRRIER.)— Sodium salt of *p*-nitro-benzene azo-*α*-naphthylamine monosulphonic acid

$$C_6H_4(NO_2)N : NC_{10}H_5NH_2SO_3Na$$

obtained by combining diazo-*p*-nitrobenzene with naphthionic acid. (1878.)— ROUSSIN AND POIRRIER.—*Brown paste*, soluble in water to a red-brown solution, in strong sulphuric acid to a magenta-red solution ; on diluting with water a brown-red precipitate is obtained ; and the same precipitate is got on adding hydrochloric acid to the aqueous solution ; caustic soda gives first a brown-red precipitate which is soluble on diluting with water.—*Dyes* wool in acid baths an orchil-red. Fairly fast to light, etc. ; much faster than Orchil.

Orseillin BB—
(BAYER.)—Sodium salt of toluene sulphonic acid azo-toluene-azo-*α*-naphthol sulphonic acid

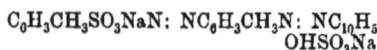

$$C_6H_3CH_3SO_3NaN: NC_6H_3CH_3N: NC_{10}H_5OHSO_3Na$$

obtained by diazotising amido-azotoluenemonosulphonic acid and combining with *α*-naphthol monosulphonic acid NW. (1883.)—*Brown powder*, soluble in water to a bluish red solution, in strong sulphuric acid with a blue colour ; on diluting with water this solution turns a crimson red. Addition of hydrochloric acid to the aqueous solution turns it yellower ; caustic soda to a red-violet solution.— *Dyes* wool in acid baths a dark orchil red, tolerably fast to washing and acids. This colour is isomeric with Crocein Scarlet 7B (BAYER), Bordeaux G (BAYER) and Ponceau 6RB (ACTIENGESELLSCHAFT).

Oxazines—
This is the name of a small group of colouring matters whose characteristic feature is that they contain the

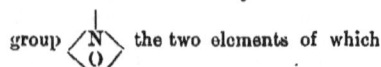

group $\diagdown\!\!\!\overset{\displaystyle N}{\underset{\displaystyle O}{}}\!\!\!\diagup$ the two elements of which

are not united together, so that they are not, to some extent, a binding group. Azo compounds, Azines, and Oxazines are related one to another to some extent. Azo

bodies contain the group—N : N—where two nitrogen atoms are joined together by two bonds and act as a binding monad group to two other groups of atoms.

Azines contain the group $\diagup\!\!\!\overset{\displaystyle N}{\underset{\displaystyle N}{|}}\!\!\!\diagdown$ in which

the two nitrogen atoms are joined together by one bond and the group acts the part of a dyad binding two dyad radicles together. In Oxazines the characteristic group acts as a dyad to one dyad radicle and as a triad to another group of radicles. The earliest known member of the series was Meldola's Blue, which is obtained by the action of nitrosodimethylaniline hydrochloride on *B*-naphthol, and is generally considered to be the chloride of dimethylphenylammonium-*B*-naphthoxazine :—

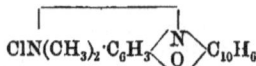

$$ClN(CH_3)_2 \cdot C_6H_3 \diagdown\!\!\!\overset{\displaystyle N}{\underset{\displaystyle O}{}}\!\!\!\diagup C_{10}H_6$$

the commercial product which is sold as New Blue, Naphthylene Blue, Fast Blue ; is not a simple colouring matter, but rather a mixture of several which have as yet not been completely separated one from the other. New Blue R is perhaps the purest of the New blues. It is the chloride of dimethylamido-naphthophen-oxazine. From the properties of these colours it may be assumed that a complete series of oxazine dyes from a pure blue to a red exist. Muscarine, which is isomeric with Meldola's Blue, and Nile Blue, which is the amido compound of the same colour, Gallocyanine and prune also belong to this group. They dye wool and silk directly in neutral or faintly acid bath, the results not being very good ; cotton they dye on a tannin-antimony mordant and give excellent results, the shades being fast to acids, alkalies and light, some of the shades resembling indigo. These colouring matters are used as a substitute for that dyestuff.

P

Pæonine—
(QUINON, MARNAS AND BONNET.)— Obtained by acting with ammonia on Aurine ; is not a definite chemical compound. PERSOZ.—*Dark-red mass, having a green lustre*, insoluble in cold, slightly soluble in boiling water to a crimson-red solution ; in alcohol to a red-brown solution ; in strong sulphuric acid to a yellow solution ; on diluting with water it turns purple-red.—*Dyes* wool and silk bright bluish reds ; not fast. Known also as Red Coralline, Aurine R.

Palatine Orange—

Ammonia salt of tetranitro-gamma-diphenol

$$(C_6H_2)_2 \begin{cases} ONH_4 \\ (NO_2)_2 \\ (NO_2)_2 \\ ONH_4 \end{cases}$$

obtained by acting on benzidin with nitric acid. (1869.) CARO.—*Brown powder*, slightly soluble in water, acids give a brown flocculent precipitate of tetranitro-gamma-diphenol. — *Dyes* wool and silk orange-yellow shades. Not now in use.

Palatine Red—

(BADISCHE.) — 1890. — *Violet - brown powder*, soluble in water to a crimson solution, in strong sulphuric acid to a blue solution ; on diluting with water this turns red. Hydrochloric acid and caustic soda added to the aqueous solution have little action.—*Dyes* wool from acid baths, bright crimson, fast to acids ; caustic soda turns it browner ; fast to light and moderately fast to washing.

Palatine Scarlet—

(BADISCHE.)— 1890. — *Maroon-coloured powder*, soluble in water to a scarlet solution, in strong sulphuric acid to a crimson solution. Hydrochloric acid has no action, caustic soda turns it a little yellower.—*Dyes* wool from acid baths bright scarlet, fast to acids, light, and alkalies; soaping discharges the colour.

Paraphenylene Blue·—

(DAHL AND CO.)—Obtained by melting *p*-diamidobenzene with hydrochlorate of amido - azo - benzene. (1888.) — *English Patent* 10,134, 1886.—*Dark blackish blue powder*, easily soluble in hot water, in sulphuric acid to a blue solution. The aqueous solution is precipitated by caustic soda, tannic acid.—*Dyes* cotton mordanted with tannin and tartar emetic, wool in slightly acid baths, deep blues, fast to acids, light, and washing.

Paris Green—

Syn. of Methyl Green.

Paris Violet—

Syn. of Methyl Violet B.

Paris Violet 6B—

Syn. of Benzyl Violet.

Patent Blue—

(MEISTER, LUCIUS.) — 1889. — *Dark blue powder*, readily soluble in water to a fine azure blue solution, in sul-

phuric acid to a brown yellow solution. Dilute hydrochloric acid added to the aqueous solution turns it green ; strong acid turns it brown-yellow ; caustic soda has little action.—*Dyes* wool from a slightly acid bath, silk from a faintly acidulated soap bath, fine greenish blue ; fast to washing ; dilute acids turn it green, strong acids brown-yellow.

Perkin's Violet—

Syn. of Mauveine.

Phenamine—

Syn. of Mauveine.

Phenanthrene Red—

(ACTIENGESELLSCHAFT.)--Sodium salt of α-naphthyl α-sulphonic acid, osazon-phenanthrenequinone

$$C_6H_4CNNHC_{10}H_6SO_3Na$$
$$C_6H_4CNNHC_{10}H_6SO_3Na$$

obtained by converting naphthionic acid into naphthyl-hydrazin-sulphonic acid, and combining with phenanthrene.(1886.) —*Brown-red powder*, slightly soluble in cold, easily soluble in hot water to a dark scarlet solution ; in strong sulphuric acid to a blue solution ; on diluting with water a yellow-brown precipitate is obtained. Hydrochloric acid gives a brown precipitate in the aqueous solution, and caustic soda a yellow-brown solution. — *Dyes* wool from an acid bath dark shades of scarlet.

Phenetol—

$$C_6H_5OC_2H_5$$

(*ethyl phenate*), made by acting with iodide of ethyl on potassium phenate ; is used in making a few dyes. From it is obtained by nitration and subsequent reduction

Phenetidine $C_6H_4O.C_2H_5NH_2$. and
Diphenetidine $C_6H_3OC_2H_5NH_2$
$$C_6H_3OC_2H_5NH_2$$

used in making benzoazurine, etc.

Phenetol Red—

Sodium salt of Phenetolazo-*B*-naphthol disulphonic acid.

$$C_6H_4OC_2H_5N : NC_{10}H_4OH(SO_3Na)_2$$

obtained by treating diazo-o-phenetol with *B*-naphthol disulphonic acid R. (1878.) *English Patent* 4,726, 1878.—Properties same as Anisol Red. Known also as Coccinin. Not now in use.

Phenicienne—

Syn. of Phenyl Brown.

Phenicin—

Syn. of Phenyl Brown.

Phenol—

$$C_6H_5OH.$$

Carbolic acid. Is found in coal tar and during the distillation of it collects in the so-called creosote oils, from which it is removed by treatment with caustic soda. It forms colourless crystals, which on exposure to air frequently turn red. Has a peculiar characteristic odour, a strong caustic action on the skin. Is only slightly soluble in water, soluble in glycerine, alcohol, etc., and in alkalies. It is a powerful disinfectant and antiseptic. It is used in the manufacture of several dyes, picric acid, etc.

Phenol Phthaleln—

Dioxyphthalophenone

$$C \begin{cases} C_6H_4OH \\ C_6H_4OH \\ C_6H_4COO \end{cases}$$

obtained by heating phthalic anhydride with phenol in the presence of chloride of tin or sulphuric acid. (1871.)—*Faintly yellow crystals,* insoluble in water, soluble in alcohol to colourless solution, in strong sulphuric acid to a pale brownish yellow solution ; on diluting with water the substance is precipitated out. Caustic soda added to the aqueous solution turns it crimson. Not used in dyeing; used as an indicator in chemical analysis. The change of colour, on the addition of alkalies, to red, or of acids from red to white, is very sharp. Also used for preparing other colours.

Phenosafranine—

(WILLIAMS, THOMAS AND DOWER.)—Paraamidophenyl - para - amidophenazonium chloride

$$NH_2C_6H_3 \diagdown \!\!\! \begin{matrix} N \\ N \end{matrix} \!\!\! \diagup \begin{matrix} C_6H_4 \\ C_6H_4NH_2 \end{matrix}$$
$$Cl$$

obtained by oxidising one molecule of p-phenylenediamine with two molecules of aniline. (1878.) O. N. WITT.—*Green, lustrous crystals,* easily soluble in water to a carmine solution, in strong sulphuric acid to a green solution which on diluting with water turns first blue, then violet, finally crimson. Dilute hydrochloric acid added to the aqueous solution turns it bluer, strong acid gives a blue solution. Caustic soda gives a red-brown precipitate.—*Dyes* cotton mordanted with tannin and tartar emetic, wool and silk in neutral baths,

fine bright crimson reds ; not fast to light, fairly so to acids, and washing. Known also as Safranine B Extra (BADISCHE).

Phenyl Brown—

Principally dinitro-phenol prepared by treating phenol with nitric and sulphuric acids. (1863.) ROTH.—*Yellow - brown powder,* very little soluble in water, alkalies turn the solution blue.—*Dyes* wool and silk a pale brown. Not now in use. Known also as Phenicienne, Phenicin, Rothein.

Phenyl Violet—

(BROOKE, SIMPSON AND SPILLER.) — Hydrochlorate of mono and diphenyl rosaniline

$$C \begin{cases} C_6H_3NH_2CH_3C_6H_5 \\ C_6H_4NH_2 \\ C_6H_4NHHCl \end{cases}$$

obtained by phenylating rosaniline. Insoluble in water, soluble in alcohol ; used to a small extent in wool dyeing, giving dull violets, fast to milling ; known also as Spirit Violet.

Phenylene Brown—

Syn. of Bismarck Brown.

Philadelphia Yellow—

Syn. of Phosphine.

Phloxin—

Sodium salt of tetrabrom tetrachlorfluorescein

$$C \begin{cases} C_6H.Br_2ONa. \\ C_6H.Br_2ONa. \\ C_6Cl_4 \ \ COO \end{cases} \!\!\! \diagdown O \diagup$$

obtained by acting with bromine upon tetrachlorfluorescein in alcoholic solution. (1882.)—*Dark-red powder,* easily soluble in water to a deep bluish-red solution, which has a weak dark-green fluorescence ; soluble in alcohol to a bluish red solution, having a dark-red fluorescence ; soluble in strong sulphuric acid to a yellow solution ; on diluting with water a red flocculent precipitate is obtained. Hydrochloric acid added to the aqueous solution gives a reddish flocculent precipitate. Caustic soda causes little change. Precipitated by solutions of lead acetate, alumina sulphate.—*Dyes* silk and wool bluish reds, bright, but not fast. Known also as Erythrosin B. (SOCY. CHEM. IND., BASLE.)

Phloxine P—

(BADISCHE.)—Alkali salt of tetrabrom-dichlorfluorescein

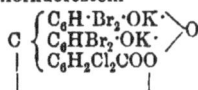

$$C \begin{cases} C_6H \cdot Br_2 \cdot OK \\ C_6HBr_2 \cdot OK \\ C_6H_2Cl_2COO \end{cases} O$$

obtained by treating dichlorfluorescein with bromine. (1875.) *English Patent* 447, 1879.—*Brown-yellow powder*, soluble in water to a scarlet solution, having a fine greenish-yellow fluorescence. Soluble in strong sulphuric acid to a brown-yellow solution, which on heating does not change; on diluting with water a pale, brownish precipitate is obtained. Hydrochloric acid added to the aqueous solution on warming gives a brown-yellow precipitate. Caustic soda turns the colour of the solution more bluish. Precipitated by solutions of lead acetate, alumina sulphate.—*Dyes* wool and silk very bright bluish scarlets without fluorescence.

Phloxine TA—

(MONNET.)—Syn. of Phloxine (MEISTER, LUCIUS).

Phosphine—

(BROOKE, SIMPSON AND SPILLER.)—The nitrate of chrysaniline (diamido-phenyl-acridine)

$$C.(C_6H_4)_2NH_2C_6H_3NH_2HNO_3.N.$$

obtained as a by-product in the preparation of magenta by the arsenic process. (1861.) NICHOLSON.—*Is a yellow or orange-yellow powder*, easily soluble in water or alcohol. Hydrochloric acid precipitates the diacid salt, nitrates precipitate the nitrate of chrysaniline from the solution, ammonia or caustic alkalies precipitate chrysaniline $C_{20}H_{17}N_3H_2O$.—*Dyes* on wool in acid bath, silk in a simple soap bath, on cotton with an alumina mordant. Also known as Aniline Orange, Chrysaniline, Philadelphia Yellow.

Picric Acid—

Trinitrophenol

$$C_6H_2OH(NO_2)_3.$$

(1842.) By LAURENT from phenol; previously, in 1771, obtained from indigo by WOULFE. Prepared by treating phenol sulphonic acid with nitric acid.—*Pale yellow needle-shaped crystals*, slightly soluble in water, and the solution has a very bitter taste; soluble in alcohol, benzol. Boiling with potassium cyanide gives a brown solution. —*Dyes* wool and silk in acid bath yellow. It is very strong: one part will colour 1,000 parts of silk. Or exposure to light

and air the colour darkens; it is not fast to washing, and it may thus be removed from the fibre. Its most useful application is in the production of mixed shades with the basic aniline colours, as with these it forms insoluble colour lakes.

Picryl Orange—

Sodium salt of trinitrophenylamido-benzene-azo-benzene sulphonic acid

$$C_6H_2(NO_2)_3NHC_6H_4N : NC_6H_4SO_3Na$$

obtained by the action of picryl chloride upon amido-azo-benzo sulphonic acid. (1882.) NOELTING.—*Orange-yellow crystals*, easily soluble in hot water to a brown-yellow solution, in strong sulphuric acid to a crimson solution; on diluting with water this turns yellow. Hydrochloric acid added to the aqueous solution precipitates the free colour acid in orange yellow flocculent mass. Caustic soda turns the aqueous solution brown.—*Dyes* wool in acid baths orange. Not now in use.

Picryl Yellow—

Sodium salt of trinitrophenylnaphthylamine sulphonic acid

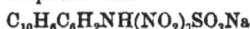

$$C_{10}H_6C_6H_2NH(NO_2)_3SO_3Na$$

obtained by treating naphthionic acid with picrylchloride. (1882.) E. NOELTING. —*Gold-yellow plates*, slightly soluble in water, acids precipitate the free acid.— *Dyes* wool in acid baths orange-yellows. Not now in use.

Pigment Brown—

(BADISCHE.)—Syn. of Soudan Brown.

Pittakal Euplttonacid—

Hexaoxymethylaurin

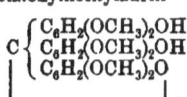

$$C \begin{cases} C_6H_2(OCH_3)_2OH \\ C_6H_2(OCH_3)_2OH \\ C_6H_2(OCH_3)_2O \end{cases}$$

obtained by the oxidation of pyrogallol and methylpyrogallolether. (1835.) REICHENBACH. — *Crystalline powder*, melting at 200deg.C., soluble in alkaline solutions with a blue colour; caustic soda gives a blue precipitate from this solution.—*Dyes* animal fibres from acid baths orange, or in ammoniacal solutions on tin mordants a blue-violet. Not now in use.

Ponceau G—

(BADISCHE.)—Sodium salt of xylene-azo-B-naphthol disulphonic acid

$$C_6H_3(CH_3)_2 N : NC_{10}H_4OH(SO_3 Na)_2$$

obtained by treating diazoxylene with

B-naphthol disulphonic acid G. (1878.)
—Has similar properties to Scarlet G
(BADISCHE) but dyes yellower shades.

Ponceau B—

(MEISTER, LUCIUS AND BRUNING.) —
Syn. of Biebrich Scarlet.

Ponceau 2 G—

(ACTIENGESELLSCHAFT, MEISTER, LUCIUS
AND BRUNING.)—Sodium salt of benzene-
azo-B-naphthol disulphonic acid R

$$C_6H_5N:NC_{10}H_4OH(SO_3Na)_2$$

obtained by treating diazobenzene with
B-naphthol disulphonic acid R. (1878.)—
Fiery-red powder, soluble in water to a
reddish yellow solution, in strong sulphuric acid to a cherry-red solution; on diluting with water a reddish yellow solution
is obtained. Hydrochloric acid added to
the aqueous solution turns it slightly
yellower, caustic soda turns it yellower.—
Dyes wool and silk in acid baths reddish
orange shades.

Ponceau 3 G—

(BADISCHE.)—Sodium salt of anisidin
sulphonic acid-azo-B-naphthol

$$C_6H_3SO_3NaOCH_3N : NC_{10}H_6OH$$

obtained by diazotising anisidine monosulphonic acid and combining with B-
naphthol. (1879.)

Ponceau 4 GB—

(ACTIENGESELLSCHAFT.)—Sodium salt of
benzene-azo-b-naphthol monosulphonic
acid G

$$C_6H_5N:NC_{10}H_6OHSO_3Na$$

obtained by combining diazobenzene with
b-naphthol monosulphonic acid C. (1878)
—*Fiery-red powder*, soluble in water to
an orange-yellow solution, in strong
sulphuric acid to an orange-yellow
solution; on adding water a yellow-brown
precipitate is formed; on adding hydrochloric acid to the aqueous solution a
yellow-brown precipitate is obtained.
Caustic soda produces a brown-yellow
solution.—*Dyes* wool in acid baths an
orange-yellow. Known also as Crocein
Orange (KALLE. BAYER AND CO.) and
Brilliant Orange (MEISTER, LUCIUS AND
BRUNING).

Ponceau GT—

(BADISCHE.)—Sodium salt of toluene-azo-
B-naphtho-disulphonic acid

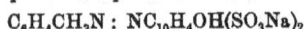

$$C_6H_4CH_3N : NC_{10}H_4OH(SO_3Na)_2$$

obtained by treating diazotoluene with B-
naphthol disulphonic acid G. (1878.)—
Red powder, soluble in water to a yellow-

ish red solution, in strong sulphuric acid
to a cherry-red solution; on diluting with
water the colour changes to orange; the
addition of hydrochloric acid to the
aqueous solution causes no change;
caustic soda turns the colour of the solution brownish.—*Dyes* wool in acid baths
a deep orange.

Ponceau RR—

(HENRIET, ROMAN AND VIGNON.)—
Sodium salt of amido-azo-benzene-azo-
B-naphthol monosulphonic acid

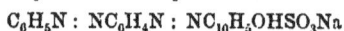

$$C_6H_5N : NC_6H_4N : NC_{10}H_5OHSO_3Na$$

obtained by diazotising amido-azo-benzene
and combining with B-naphthol monosulphonic acid B and S. (1879.) *English
Patent* 5,003, 1879.—*Brown powder*,
soluble in water to a cherry-red solution,
and in strong sulphuric acid to a blue
solution; on addition of water a brown
flocculent precipitate first forms; this, on
adding more water, dissolves. Hydrochloric acid added to the aqueous solution
gives a brown precipitate which is soluble
in water; caustic soda gives a violet
precipitate soluble in water.—*Dyes* wool
red shades in acid baths.

Ponceau 2R—

(ACTIENGESELLSCHAFT, MEISTER, LUCIUS
AND BRUNING.)—Sodium salt of xylene-
azo-B-naphthol disulphonic acid

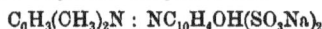

$$C_6H_3(CH_3)_2N : NC_{10}H_4OH(SO_3Na)_2$$

obtained from diazoxylene by treating
with B-naphthol disulphonic acid R.
(1878.)—Has similar properties to Scarlet
G (BADISCHE). Known also as Xylidine
Red and Xylidine Scarlet.

Ponceau 3R—

(ACTIENGESELLSCHAFT AND MEISTER,
LUCIUS AND BRUNING.)—Sodium salt of p-
cumidin-azo-B-naphthol disulphonic acid

$$C_6H_2(CH_3)_3N : NC_{10}H_4OH(SO_3Na)_2$$

obtained by treating diazo p-cumene with
B-naphthol disulphonic acid R. (1878.)—
Dark-red powder, soluble in water to
a cherry-red solution, and in strong
sulphuric acid to a cherry-red solution;
on adding water to this there is no change.
Acids do not change the colour of the
aqueous solution; caustic soda gives a
yellow precipitate.—*Dyes* wool and silk
bluish reds in acid baths, the shades are
fairly fast to acids and light. Known also
as Cumidine Red and Cumidine Scarlet.

Ponceau 3RB—

(ACTIENGESELLSCHAFT.)—Syn. of Biebrich Scarlet.

Ponceau 3R—

(MEISTER, LUCIUS AND BRUNING.)—Sodium salt of ethyl-dimethyl-benzene-azo-B-naphthol disulphonic acid

$$C_6H_2(C_2H_5)(CH_3)_2N : NC_{10}H_4OH(SO_3Na)_2$$

obtained by treating diazoethyl-dimethyl-benzene with B-naphthol disulphonic acid R. (1878.)—Same properties as Ponceau 3R from Cumidine.

Ponceau 4R—

(ACTIENGESELLSCHAFT.)—This has the same composition as Ponceau 3R, but is made from chemically pure Cumidine.

Ponceau 4RB—

(ACTIENGESELLSCHAFT.)—Syn. of Crocein Scarlet 3B (BAYER).

Ponceau 5R—

(MEISTER, LUCIUS AND BRUNING.)—Sodium salt of amido-azo-benzene-B-naphthol trisulphonic acid

$$C_6H_5N : NC_6H_4N : NC_{10}H_3OH (SO_3Na)_3$$

obtained by diazotising amido-azo-benzene and combining with B-naphthol trisulphonic acid. (1882.)—*Brown powder*, soluble in water to a bluish red solution, and in strong sulphuric acid to a purple solution ; on diluting with water this first turns blue, then red, no precipitate being formed; on adding hydrochloric acid to the aqueous solution it turns brown.—*Dyes* wool and silk in acid baths fine shades of crimson, and cotton similar shades in a boiling bath. Also known as Errythine (BADISCHE).

Ponceau 6R—

(MEISTER, LUCIUS AND BRUNING. BADISCHE.)—Sodium salt of naphthionic acid-azo-B-naphthol trisulphonic acid

$$C_{10}H_6SO_3NaN : NC_{10}H_3OH(SO_3Na)_3$$

obtained by diazotising naphthionic acid and combining it with B-naphthol trisulphonic acid. (1882.) — *Brown Powder*, soluble in water to a crimson solution, in strong sulphuric acid to a violet solution ; on diluting with water this turns crimson. Addition of hydrochloric acid to the aqueous solution does not alter it, while caustic soda turns it a red-brown.—*Dyes* wool in acid baths a bright bluish-red, fast.

Ponceau 6RB—

(ACTIENGESELLSCHAFT.)—Syn. of Crocein Scarlet 7B (BAYER).

Ponceau RT—

(BADISCHE.)—Sodium salt of toluene-azo-B-naphthol disulphonic acid. Isomeric with Ponceau GT.

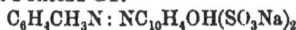

$$C_6H_4CH_3N : NC_{10}H_4OH(SO_3Na)_2$$

obtained by treating diazotoluene with B-naphthol disulphonic acid R. (1878.)—Properties same as Ponceau GT.—*Dyes* wool and silk in acid baths a reddish orange.

Ponceau S Extra—

(ACTIENGESELLSCHAFT.) — Sodium salt of benzene sulphonic acid-azo-benzene sulphonic acid-azo-B-naphthol disulphonic acid

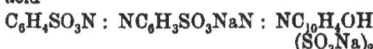

$$C_6H_4SO_3N : NC_6H_3SO_3NaN : NC_{10}H_4OH(SO_3Na)_2$$

obtained by diazotising amido-azo-benzene disulphonic acid and combining with B-naphthol disulphonic acid R. (1880.)—*Brown powder*, soluble in water to a bluish red solution ; in strong sulphuric acid to a blue solution ; on diluting this with water a scarlet solution is obtained ; addition of hydrochloric acid does not alter the aqueous solution ; caustic soda gives a violet precipitate soluble on the addition of much water.—*Dyes* wool in acid baths fine bluish reds, fast to acids, washing and light. Also known as Fast Ponceau 2B (BADISCHE).

Ponceau SS Extra—

(ACTIENGESELLSCHAFT.)—Sodium salt of amido-azo-benzene-azo-B-naphthol disulphonic acid

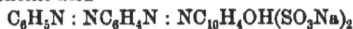

$$C_6H_5N : NC_6H_4N : NC_{10}H_4OH(SO_3Na)_2$$

obtained by diazotising amido-azo-benzene and combining with B-naphthol disulphonic acid R. (1879.) *English Patent* 5,003, 1879.—*Brown Powder*, soluble in water to a bluish red solution ; in strong sulphuric acid to a violet solution ; on diluting with water a violet precipitate is obtained, this is also got by adding hydrochloric acid to the aqueous solution ; on adding caustic soda the aqueous solution it turns red-violet.— *Dyes* wool and silk in acid baths, and cotton in a boiling bath, fine bluish reds, fast to light, acids and washing.

Primrose—

(BADISCHE.)—Potassium salt of tetrabromfluorescein ethyl ether

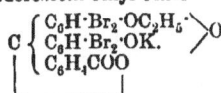

$$C \begin{cases} C_6H \cdot Br_2 \cdot OC_2H_5 \cdot \\ C_6H \cdot Br_2 \cdot OK. \\ C_6H_4COO \end{cases} O$$

obtained by ethylating eosin. (1874.

L

CARO.—*Brown crystalline powder with a greenish lustre*, very slightly soluble in cold water, soluble in hot water to a scarlet-red solution; the solution has a weak greenish yellow fluorescence. Soluble in strong sulphuric acid to a yellow solution; on heating, bromine is evolved; on diluting with water a brown-yellow precipitate is obtained; hydrochloric acid added to the aqueous solution gives a yellow-brown precipitate; caustic soda gives a brownish-yellow precipitate. It is also precipitated by solutions of lead acetate, alumina sulphate.—*Dyes* wool and silk from a weak acid bath, bright scarlets having a weak flourescence. The shades are fast to acids and washing but not to light. Also known as Spirit Eosine, Ethyl Eosine, Eosin S, Eosin BB, Rose J B.

Primrose Soluble—

(DURAND.)—Syn. of Erythrosine.

Primula—

Syn. of Hofmann's Violet.

Primuline—

(BROOKE, SIMPSON AND SPILLER.)—A yellow colouring matter obtained by acting with sulphur on paratoluidine and sulphonating the product. It is a combination of the sulphonic acids of two bases, the so-called " Primuline base," which is probably amidobenzenyl-amidothiocresol.

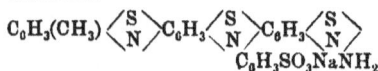

$$C_6H_3(CH_3)\left\langle\begin{matrix}S\\N\end{matrix}\right\rangle C_6H_3\left\langle\begin{matrix}S\\N\end{matrix}\right\rangle C_6H_3\left\langle\begin{matrix}S\\N\end{matrix}\right\rangle$$
$$C_6H_3SO_3NaNH_2$$

and dehydrothiotoluidine

$$C_6H_3(CH_3)\left\langle\begin{matrix}S\\N\end{matrix}\right\rangle C_6H_4(NH_2)$$

which is present in small quantities only. (1887.) GREEN. GREEN. *Jour. Socy. Chem. Ind.*, 1888, p. 179. *Chem. Soc. Jour.*, 1889, p. 227.—*A lemon-yellow powder*, soluble in water to a yellow solution, in strong sulphuric acid to a yellow solution with a strong green fluorescence.—*Dyes* cotton, wool, or silk from a salt bath greenish yellow, moderately fast to acids, reddened by alkalies and by soaping; not fast to light.

The importance of primuline as a dyestuff lies in the fact that it can serve by the process of diazotising and developing for the production of other and faster colours; this can take place on fibre dyed with primuline, and in this way can be produced what were called by Green " Ingrain Colours," a term which might well be used to indicate colours developed on fibre from colour bases in the same way. In the case of primuline the cotton or silk is dyed with the dyestuff in the usual way, then it is passed into an acidulated bath of sodium nitrite for 15 minutes, then taken, rinsed slightly, and passed into a developing bath. Upon the nature of this bath depends the colour which will be formed : if it consists of an alkaline solution of *b*-naphthol a bright red is obtained ; if of an alkaline solution of *a*-naphthol a dark red is obtained ; a solution of Schaffer's acid gives scarlet; of naphthylamine hydrochlorate a grenat; of naphthylamine other a blue; of resorcin an orange ; of phenol a yellow ; of phenylene diamine a brown. As a rule these colours are fast to acids, alkalies, and washing, and are fairly fast to light. For their properties the paper by Green, already quoted, may be referred to. Also known as Auroline, Carnotine, Polychromine, Throchromogen, Yellow PR, etc.

Propiolic Acid—

(BADISCHE.)—Ortho-nitro-phenyl propiolic acid

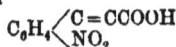

$$C_6H_4\left\langle\begin{matrix}C=COOH\\NO_2\end{matrix}\right.$$

obtained by treating cinnamic acid dibromide with caustic soda. (1880.) BAYER.—*Yellowish white paste, or tabular crystals*, soluble in water and alcohol ; on heating with glucose and caustic soda is converted into white indigotin. Used to a very small extent in calico printing for the production of indigo blues. Propiolic acid, as the source of indigo, has practically gone out of use; it is more costly than natural indigo, and there are also some practical difficulties which prevent its successful application.

Prune—

(KERN AND SANDERS.)—Methylether of gallocyanine

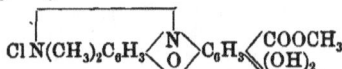

$$Cl\ N(CH_3)_2C_6H_3\left\langle\begin{matrix}N\\O\end{matrix}\right\rangle C_6H_3\left\langle\begin{matrix}COOCH_3\\(OH)_2\end{matrix}\right.$$

obtained by acting with nitrate of nitrosodimethylaniline on methylether of gallic acid. (1886.) KERN.—*Dark-brown powder*, easily soluble in water and alcohol to a blue-violet solution; soluble in strong sulphuric acid to a bright-blue solution ; on diluting with water this turns a magenta colour. Hydrochloric acid added to the aqueous solution turns it crimson ; caustic soda gives first a brown precipitate then a violet solution.—*Dyes* wool or cotton mordanted with chrome a blue-violet ; dyes also cotton mordanted with tannin. The shades are fairly fast to light, acids and washing.

Purpurin—

Trihydroxyanthraquinone

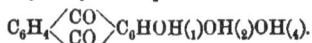

$$C_6H_4\diagup{CO}\diagdown{CO}\diagup C_6HOH(_1)OH(_2)OH(_4).$$

This is found along with alizarine in the root of the madder ; is found in the crude alizarine melt. Purpurin crystallises in *orange needles*, melts at 253deg. C., but sublimes at 150deg, C. ; slightly soluble in water to a deep yellow solution, in alkalies with a purplish red colour.—*Dyes* fibres similar to alizarine and anthrapurpurin ; the reds are much yellower, but the browns with chrome are deeper than with the other two bodies. Known as Alizarine No. 6 (MEISTER).(See ALIZARINE.)

Purpurin—

An old name for Mauveine.

Pyronine G—

(LEONHARDT AND CO.)

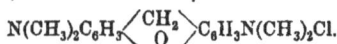

$$N(CH_3)_2C_6H_3\diagup{CH_2}\diagdown{O}\diagup C_6H_3N(CH_3)_2Cl.$$

obtained by the action of benzaldehyde on dimethylamidophenol. (1890.) *English Patent* 13,217, 1889. — *Brown powder*, which is soluble in water to a red solution ; slightly soluble in alcohol to a bluish pink solution having a slight fluorescence ; in acetic acid a solution has a bluish red colour and an orange fluorescence ; strong sulphuric acid forms a yellow solution, which on diluting with water turns orange. Hydrochloric acid has no action on the aqueous solution ; caustic soda throws down a violet red precipitate. — *Dyes* cotton which has been mordanted with tannin : wool and silk from neutral baths fine bluish pinks of great brightness ; acids turn the shade more orange ; caustic soda gradually discharges the colour ; not quite fast to strong soaping ; fairly fast to light.

Pyrosin B—

(MONNET.)—Syn. of Erythrosine.

Pyrosine J—

(MONNET.)—Syn. of Erythrosine G.

Pyrosine R—

(MEISTER, LUCIUS.)—Is a mixture of the alkali salts of triod and tetriod fluorescein ; is, therefore, a mixture of Erythrosine and Erythrosine G.

Pyrotin—

(DAHL.)—Sodium salt of β-naphthalene-sulphonic acid-azo-α-naphthol monosulphonic acid

$$C_{10}H_6SO_3NaN : NC_{10}H_6OHSO_3Na$$

obtained by diazotising β-naphthylamine monosulphonic acid D, and combining with α-naphthol monosulphonic acid NW. (1884.) *English Patent* 7,713, 1884.— *Brown-red solution* ; in strong sulphuric acid to a crimson solution ; on diluting with water this turns red ; adding hydrochloric acid to the aqueous solution turns it a little more bluish, caustic soda more scarlet.—*Dyes* wool in acid baths bright scarlet reds, fairly fast to acids, etc. This colour is isomeric with Double Scarlet S Extra (ACTIENGESELLSCHAFT).

Q

Quinoline Blue—

(GEIGY.)—Syn. of Cyanine.

Quinoline Green—

(BADISCHE.)—Hydrochlorate of tetramethyldiamido-diphenylquinolyl-carbinol

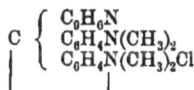

$$C\begin{cases} C_9H_6N \\ C_6H_4N(CH_3)_2 \\ C_6H_4N(CH_3)_2Cl \end{cases}$$

obtained by acting with quinoline upon tetramethyldiamido benzophenone chloride. (1883.) CARO AND KERN. *English Patent* 5,003, 1884 ; 11,159, 1884. Not now in use.

Quinoline Red—

(ACTIENGESELLSCHAFT.)

$$ClC\begin{cases} C_6H_4 \\ CH_2C_9H_6N \\ C_9H_6N \end{cases}$$

acting with benzotrichloride upon a mixture of quinaldine and isoquinoline in presence of chloride of zinc. (1882.) — *Dark brown-red needle-shaped crystals*, with a bronzy lustre, insoluble in cold, slightly soluble in hot, water ; soluble in alcohol to a red solution, having a yellowish red flourescence ; in strong sulphuric acid to a colourless solution ; on diluting with water this turns red. Not used in dyeing ; used in photography to a small extent.

Quinoline Yellow—

Sodium salt of the sulphonic acid of quinaldylenphthalid

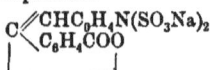

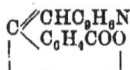

$$C \Big\backslash{}^{CHC_9H_4N(SO_3Na)_2}_{C_6H_4COO}$$

obtained by acting upon spirit soluble quinoline yellow with strong sulphuric acid. (1882.) — *Lemon-yellow powder,* easily soluble in water to a greenish yellow solution; soluble in alcohol; soluble in strong sulphuric acid to a red solution, which on diluting with water turns yellow ; hydrochloric acid turns the colour of the aqueous solution paler, caustic soda turns it darker.—*Dyes* wool and silk from acid baths fine greenish yellow shades, very fast to acids, washing and light.

Quinoline Yellow (*Spirit Soluble*)—

Quinaldylenphthalid

$$C \Big\backslash{}^{CHC_9H_9N}_{C_9H_4COO}$$

obtained by heating quinaldine with phthalic anhydride and chloride of zinc. (1882.)—*Yellow powder,* insoluble in water; soluble in alcohol to a yellow solution; soluble in strong sulphuric acid to a yellowish red solution ; on diluting with water yellow flocculent precipitate is obtained. Used for colouring varnishes, fats.

R

Red B—

(BADISCHE.)—Syn. of Soudan II.

Red C—

(BADISCHE.)—Syn. of Fast Red B.

Red Coralline—

Syn. of Pæonine.

Red Violet 5R—

Syn. of Hofman's Violet.

Red Violet 5RS—

(BADISCHE.)—Sodium salt of ethylrosaniline sulphonic acid

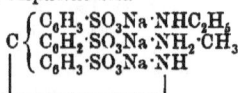

$$C \Big\{{}^{C_6H_3 \cdot SO_3Na \cdot NHC_2H_5}_{C_6H_2 \cdot SO_3Na \cdot NH_2 \cdot CH_3}_{C_6H_3 \cdot SO_3Na \cdot NH}$$

obtained by acting upon ethylrosaniline

with sulphuric acid. (1877.) — *English Patent* 3,731, 1877.—*Brown-violet lumps with a metallic lustre,* soluble in water to a crimson-red solution ; in strong sulphuric acid to a yellow solution, which becomes crimson on diluting with water ; adding hydrochloric acid to the aqueous solution has little action ; caustic soda turns it brown-yellow.—*Dyes* wool and silk in acid baths bluish reds, not very fast to light or washing.

Regina Purple—

(BROOKE, SIMPSON AND SPILLER.) — Usually the acetate of *o*-tolyl-*p*-rosaniline

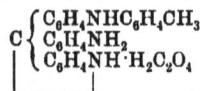

$$C \Big\{{}^{C_6H_4NHC_6H_4CH_3}_{C_6H_4NH_2}_{C_6H_4NH \cdot H_2C_2O_4}$$

obtained by acting with the oil from the arsenic-magenta process upon a mixture of rosaniline and acetic acid. (1860.) *English Patent,* January 12th, 1861. Introduced by Simpson, Maule and Nicholson.—*Green powder,* easily soluble in water to a red-violet solution ; in strong sulphuric acid to a brown solution ; on diluting this with water it becomes paler ; hydrochloric acid added to the aqueous solution turns it brown ; on diluting with water this turns blue ; caustic soda gives a brown precipitate.—*Dyes* wool and silk in neutral bath ; cotton requires a mordant of tannin and tartar emetic ; red-violet shades are obtained which are fairly fast to light or acids. Known also as Regina Violet and Imperial Violet.

Regina Violet—

Syn. of Regina Purple.

Regina Violet (*Spirit Soluble*)—

(ACTIENGESELLSCHAFT.) — Bye - product of the nitrobenzol process of magenta making. Probably the hydrochlorate of diphenylrosaniline. — *Green granules,* slightly soluble in water, easily in alcohol ; in strong sulphuric acid to a green solution ; on adding water the sulphate of the base is thrown down. Used for making varnishes.

Regina Violet (*Water Soluble*)—

(ACTIENGESELLSCHAFT.)—Sodium salt of the sulphonic acid of Regina Violet spirit soluble.—*Bronzy lustrous lumps,* soluble in water to a violet solution ; caustic soda nearly discharges the colour ; acids precipitate the free colour acid.—*Dyes* wool and silk from acid baths reddish violet shades.

Resorcin Blue—

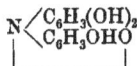

(KERN AND SANDOZ.)—Syn. of Fluorescent Blue.

Resorcin Blue—

$$N \left\langle \begin{matrix} C_6H_3(OH)_2 \\ C_6H_3OHO \end{matrix} \right.$$

obtained by heating resorcin with sodium nitrite. (1884.)—*Dark blue-violet lustrous mass*, soluble in water to a blue-violet solution ; in alcohol to a blue solution, having a dark-green fluorescence. Hydrochloric acid gives a dirty-red precipitate. Used principally as an indicator in analysis. Known also as Lakmoid.

Resorcin Brown—

(ACTIENGESELLSCHAFT.) — Sodium salt of xylene-azo-sulphanilic acid-azo-resorcin

$$C_6H_3(CH_3)_2N : NC_6H_2(OH)_2N : C_6H_4SO_3Na$$

obtained by treating diazoxylene with sulphanilic acid-azo-resorcin (Tropæolin O). (1881.)—*Brown powder*, soluble with a brown colour in water, and in strong sulphuric acid to a brown solution ; on diluting with water a brown precipitate is obtained, and on adding hydrochloric acid to the aqueous solution the same precipitate also forms; caustic soda causes little alteration in the aqueous solution. —*Dyes* wool and silk shades of reddish brown in acid baths.

Resorcin Green—

(DURAND AND HUGUENIN.)—Dinitroso-resorcin

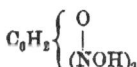

$$C_6H_2 \left\{ \begin{matrix} O \\ | \\ (NOH)_2 \end{matrix} \right.$$

obtained by treating resorcin with nitrous acid. (1875.)—*A dirty-green paste*, sparingly soluble in cold water, more freely on boiling ; soluble in dilute caustic soda. Applied to cotton with tannin and iron mordant; on wool with copperas and tartar. Gives myrtle-green shades, which are very fast. Sold also as Solid Green (BADISCHE) and Fast Myrtle. Mostly used in calico printing.

Resorcin Violet—

Tetramethyl - diamido-dioxy-triphenyl-carbinol

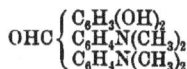

$$OHC \left\{ \begin{matrix} C_6H_3(OH)_2 \\ C_6H_4N(CH_3)_2 \\ C_6H_4N(CH_3)_2 \end{matrix} \right.$$

obtained by acting with resorcin upon tetramethyl-diamido-benzophenone chloride., (1883.) CARO.—Not now in use.

Resorcin Yellow—

(WILLIAMS, THOMAS AND DOWER.) — Sodium salt of sulphanilic acid-azo-resorcin

$$C_6H_4SO_3NaN : NC_6H_3(OH)_2$$

obtained by diazotising sulphanilic acid and combining with resorcin. (1876.) WITT. *Chem. Socy. Journ.*, 1879, p.183.— *Brown powder*, soluble in water with a reddish yellow colour; in strong sulphuric acid with a yellow colour ; on diluting with water this turns reddish yellow ; no change is produced on adding hydrochloric acid to the aqueous solution ; caustic soda turns it reddish brown. —*Dyes* wool in acid baths a reddish yellow, not very fast. Also known as Tropæolin O (original name), Tropæolin R, Chrysoin, Chryseolin, Yellow T (SOCY. CHEM. IND., BASLE), Gold-yellow (BAYER).

Rhodamine—

(BADISCHE.)—Hydrochlorate of the phthalein of diethyl-*meta*-amido-phenol

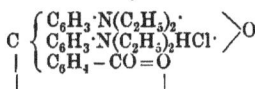

$$C \left\{ \begin{matrix} C_6H_3 \cdot N(C_2H_5)_2 \cdot \\ C_6H_3 \cdot N(C_2H_5)_2HCl \cdot \\ C_6H_4 - CO = O \end{matrix} \right\rangle O$$

made by melting phthalic anhydride with diethyl - *meta* - amido-phenol. (1887.) *English Patent.*—*Green crystals or reddish violet powder*, easily soluble in water to a crimson solution, which when dilute is fluorescent, soluble in strong sulphuric acid to a yellow-brown solution with evolution of hydrochloric acid ; on diluting with water the colour of the solution changes through scarlet-red to bluish red. Soluble in alcohol to a crimson solution, which if dilute has a strong flourescence ; on heating this disappears but returns on cooling ; hydrochloric acid added to the aqueous solution gives green crystalline precipitate of hydrochlorate ; on further addition a scarlet solution is obtained ; caustic soda gives a pale rose-red precipitate of the base, which is soluble in ether and benzol to colourless solutions.—*Dyes* wool and silk from weak acid baths fine bluish red shades, having a strong greenish fluorescence, very fast to light, acids and washing,—*Dyes* tannin mordanted cotton fine violet-reds without fluorescence, also oiled cotton with fluorescence.

Rhodamine S—

(SOCY. CHEM. IND., BASLE.)—Obtained by heating diethyl-meta-amido-phenol with succinic acid and chloride of zinc. (1890.) *English Patent 2,635, 1889.* —*Dark grey crystalline powder*, soluble in water to a crimson solution, in alcohol to a scarlet solution with an orange fluorescence ; strong sulphuric acid dissolves it with an orange colour, the solution having a light yellow fluorescence ; hydrochloric acid added to the aqueous solution has no action ; caustic soda decolourises it, giving a colourless precipitate.—*Dyes* wool and silk in acetic acid bath, cotton in a neutral bath, fine brilliant pink shades, turned scarlet with acids ; caustic soda decolourises it ; not fast to soaping or light.

Roccellin—
Syn. of Fast Red A.

Rosamine—
(WILLIAMS BROS. AND CO.)—1888.— *Dark carmine-red paste,* slightly soluble in water to a carmine solution, freely soluble in alkaline solutions ; acids give a dark violet-blue flocculent precipitate.— *Dyes* cotton from an alkaline bath fine bluish reds, fast to soap and alkalies ; acids turn it dark purple, not fast to light and air.

Rosanaphthylamine—
Syn. of Magdala Red.

Rosaniline Colours—
Magenta, the first red colouring matter obtained from coal tar, is the salt of a base known as rosaniline, which has the formula

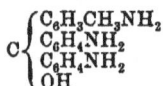

$$C \begin{cases} C_6H_3CH_3NH_2 \\ C_6H_4NH_2 \\ C_6H_4NH_2 \\ OH \end{cases}$$

This base is supposed to be derived from methane, marsh gas CH_4 by the substitution of phenyl C_6H_5 for two of its hydrogen atoms, tolyl for a third, and hydroxyl for the fourth atom ; there is, therefore, as in methane, an atom of carbon which binds the whole molecule together. Another base which is also found in commercial magenta is pararosaniline

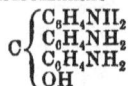

$$C \begin{cases} C_6H_4NH_2 \\ C_6H_4NH_2 \\ C_6H_4NH_2 \\ OH \end{cases}$$

This, it will be seen, is very similar to rosaniline. These bases are supposed to be derived from a body known as triphenylmethane

$$C \begin{cases} C_6H_5 \\ C_6H_5 \\ C_6H_5 \\ H \end{cases}$$

This is the parent substance not only of

magenta but of Malachite Green and other colours of this group, hence these colours are sometimes classed as the triphenylmethane group, and one method of obtaining them is to build them up from this body; for instance, when it is treated with nitric acid it yields trinitrotriphenylmethane; this, on oxidation with chromic acid, gives trinitrotriphenyl-carbinol, which, on reduction with acetic acid and zinc, gives pararosaniline ; the following formulæ show these successive stages

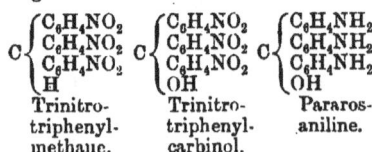

$$C \begin{cases} C_6H_4NO_2 \\ C_6H_4NO_2 \\ C_6H_4NO_2 \\ H \end{cases} \quad C \begin{cases} C_6H_4NO_2 \\ C_6H_4NO_2 \\ C_6H_4NO_2 \\ OH \end{cases} \quad C \begin{cases} C_6H_4NH_2 \\ C_6H_4NH_2 \\ C_6H_4NH_2 \\ OH \end{cases}$$

Trinitro-   Trinitro-   Pararos-
triphenyl-   triphenyl-   aniline.
methane.   carbinol.

The bases pararosaniline and rosaniline are in themselves colourless, and it is only when they are combined with acetic, hydrochloric, or other acids that they form colouring matters ; during this combination water is eliminated, thus

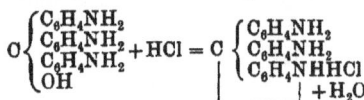

$$C \begin{cases} C_6H_4NH_2 \\ C_6H_4NH_2 \\ C_6H_4NH_2 \\ OH \end{cases} + HCl = C \begin{cases} C_6H_4NH_2 \\ C_6H_4NH_2 \\ C_6H_4NHHCl \\ \end{cases} + H_2O$$

Pararosaniline.   Magenta.

There is thus formed a mono-acid salt which is highly coloured, but di and tri acid salts can also be formed; these are generally yellow or brown, and are unstable salts, being decomposed by water reforming the mono-acid salts ; this fact explains the reason why, when fibres dyed with magenta which have been turned brown by acids have their original colour restored on washing ; the same thing happens in aqueous solutions of these dyes. By reduction the oxygen present in the "colour bases" is eliminated, and colourless bodies are formed, known as "leuco bases"; thus pararosaniline yields paraleucaniline

$$C \begin{cases} C_6H_4NH_2 \\ C_6H_4NH_2 \\ C_6H_4NH_2 \\ H \end{cases}$$

These "leuco bases" can combine with acids, but they yield colourless salts ; every colouring matter of the rosaniline group has a corresponding leuco base ; by oxidation these bases are transformed into the colour bases. The amido hydrogen atoms in these bases are capable of being replaced by methyl CH_3, ethyl C_2H_5, benzyl C_7H_7, when violet colouring matters are formed ; on the other hand, if they are replaced by phenyl C_6H_5 blue colours are produced ; the shade of the colouring matter depends upon the number of hydrogen atoms which have been so replaced ; there are six atoms of hydrogen thus replaceable, so there is possible a range of seven primary colours, from magenta on one hand to hexaphenyl rosaniline, the purest blue of the series, or from magenta to hexamethyl rosaniline, the bluest violet possible.

The diamido compound of triphenylmethane yields on oxidation the colour base

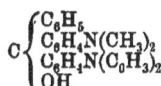

$$C\begin{cases} C_6H_5 \\ C_6H_4N(CH_3)_2 \\ C_6H_4N(C_9H_3)_2 \\ OH \end{cases}$$

the base of Malachite Green and Brilliant Green.

The colouring matters of this group can be sulphonated, yielding such colouring matters as Acid Magenta, Acid Green, Acid Mauve, Acid Violet, etc.

The rosaniline colours are what are called basic colours, *i.e.*, they have basic properties, and require an acid mordant to fix them on the cotton fibre. On cotton they are dyed by first mordanting the fibre with tannin and tartar emetic, and then dyeing in a bath of the colouring matter ; in printing, a colour is made of gum thickening, tannic acid, acetic acid, and colouring matter ; the object of adding the acetic acid is to keep the tannin-colour lake which is formed in solution. After printing, the goods are steamed, and then passed through a bath of tartar emetic which fixes the colour. This method applies to the greens, violets, reds, and the soluble blues of this group. The acid colours cannot be dyed on cotton. On wool the colours are applied in a neutral bath, containing some Glauber's salt; these colours will not dye wool or silk in an acid bath. On silk they are dyed in a bath of boiled-off liquor. The acid colours are dyed on wool and silk in an acid bath. See also METHODS OF DYEING AND PRINTING.

Roseazurine—
(BAYER.)—

Tetrazo- ⟨B-naphthylamine sodium
dianisol⟩ monosulphonate Br.
⟨B-naphthylamine sodium
monosulphonate Br.

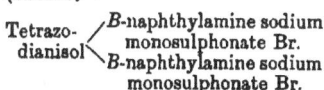

(1885).—*Brown-red powder*, soluble in water to a crimson-red solution, in strong sulphuric acid to a green-blue solution ; on diluting with water a violet-blue pre-

cipitate is obtained. Hydrochloric acid added to the aqueous solution gives a violet precipitate. Caustic soda gives a crimson solution.—*Dyes* cotton from a soap bath bluish reds. This colour is not now in use.

Roseazurine G—
(BAYER.)—

Tetrazoditolyl⟨ B-naphthylamine
sodium monosul-
phonate (*delta*)
ethyl-B-naphthyl-
amine sodium mono-
sulphonate (*delta*)

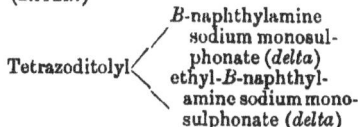

(1886.) *English Patent* 17,083, 1886.— *Dark red powder*, slightly soluble in water to a crimson solution, in strong sulphuric acid to a blue solution ; on diluting with water a red-violet precipitate is obtained. Acids added to the aqueous solution give a brown-red precipitate. Caustic soda has little action.—*Dyes* cotton from a soap bath light shades of rosy red, faster to light and acids than Congo. *Dyes* wool rather poor pale shades. *Dyes* silk light shades of rosy pink from a neutral soap bath.

Roseazurine B—
(BAYER.)—

Tretrazoditolyl⟨ ethyl-b-naphthylamine
sodium monosul-
phonate (*delta*)
ethyl-b-naphthylamine
sodium monosul-
phonate (*delta*)

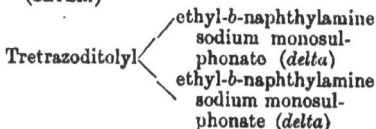

(1886.) *English Patent* 17,083, 1886.— *Brown-red powder*, slightly soluble in water to a crimson solution, in strong sulphuric acid to a blue solution ; on diluting with water a violet precipitate is obtained. Acids throw down the free colour acid from the aqueous solution as a dark crimson-red precipitate ; caustic soda has no action.—*Dyes* cotton from a soap bath a crimson-red with a bluish tint, very fine and fairly fast. Does not dye wool well. *Dyes* silk from a neutral soap bath good shades of crimson-red.

Roseazurine BB—
(BAYER.)—

Benzidin sul-⟨B-naphthylamine sodium
phontetrazo⟩ monosulphonate Br.
B-naphthylamine sodium
monosulphonate Br.

$$C_6H_3\cdot N : NC_{10}H_5NH_2SO_3Na$$
$$\Big\rangle SO_2$$
$$C_6H_3N : NC_{10}H_5NH_2SO_3Na$$

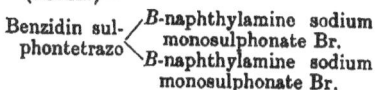

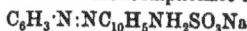

(1885.)—*Dark brown powder*, easily soluble

in water to a crimson-red solution, in strong sulphuric acid to a reddish violet solution ; on diluting with water a brown-red precipitate is thrown down. Hydrochloric acid added to the aqueous solution gives a brown-red precipitate ; caustic soda gives a blue precipitate.—*Dyes* cotton in a soap bath reddish violet shades, moderately fast to light, acids and washing ; blued by alkalies.

Rose B—

(SOCY. CHEM. IND., BASLE.)—Syn. of Erythrosine.

Rose Bengal—

Alkali salt of tetriod-dichlor-fluorescein

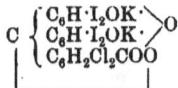

$$C \begin{cases} C_6H \cdot I_2OK \cdot \\ C_6H \cdot I_2OK \cdot \\ C_6H_2Cl_2COO \end{cases} O$$

obtained by acting with iodine upon dichlor-fluorescein. (1875.)—*Brown powder*, easily soluble in water to a deep scarlet solution, which is little fluorescent ; in strong sulphuric acid to a brown-yellow solution, which, on warming, evolves iodine ; on diluting with water a brown-red flocculent precipitate is obtained. Hydrochloric acid added to the aqueous solution gives a brown-red precipitate ; caustic soda causes little alteration in the colour ; precipitated by solutions of lead acetate, alumina sulphate.—*Dyes* wool and silk from weak acid baths fine bluish red shades, fast to acids and washing, not to light.

Rose Bengal B—

(SOCY. CHEM. IND., BASLE.)—Potassium salt of tetraiod-tetrachlor-fluorescein

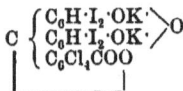

$$C \begin{cases} C_6H \cdot I_2 \cdot OK \cdot \\ C_6H \cdot I_2 \cdot OK \cdot \\ C_6Cl_4COO \end{cases} O$$

obtained by acting with iodine on tetrachlor-fluorescein. (1882.) — *Brown-red powder*, soluble in water to a bluish red solution, which has no fluorescence ; soluble in strong sulphuric acid to a brown solution ; on diluting with water flesh-red precipitate is obtained ; hydrochloric acid gives a flesh-red precipitate in the aqueous solution ; caustic soda has little action ; precipitated by solutions of lead acetate, alumina sulphate. — *Dyes* wool and silk from weak acid baths bright bluish reds.

Rose JB—

Syn. of Primrose.

Rose Bengal N—

(CASSELLA.)—Syn. of Rose Bengal.

Rosein—

Syn. of Magenta, generally given to the acetate variety.

Rosolane—

(POIRRIER).—Syn. of Mauveine.

Rosophenoline—

(CHARLES LOWE AND CO.)—A basic red colouring matter, obtained by the action of alcoholic ammonia on aurin in the presence of benzoic acid. *English Patent* No. 5,554, 1882.—This colouring matter never came into use.

Rothein—

Syn. of Phenyl Brown.

Roxamine—

(DURAND AND HUGUENIN.)—*A*-naphthylene sodium monosulphonate-azo-dioxy-naphthaline

$$C_{10}H_6SO_3NaN:NC_{10}(OH)_2$$

(1889.)—Dark brick-red powder, soluble in water to a scarlet solution, in strong sulphuric acid to a violet solution ; on diluting with water the scarlet colour is restored. Caustic soda turns the colour of the aqueous solution bluer ; hydrochloric acid has no action.—*Dyes* wool from an acid bath fine shade of scarlet, fast to light and acids, not fast to soap.

Rubeosin—

A colouring matter obtained by the action of nitric acid upon aureosin. Not now in use.—*Brown-yellow powder*, insoluble in water, soluble in alcohol to a brownish yellow solution having a green fluorescence ; soluble in alkalies to a red solution, and in strong sulphuric acid to a yellow solution, which, on diluting with water, gives a yellow precipitate.

Rubidin—

Syn. of Fast Red A.

Rubin—

Syn. of Magenta.

Rubin S—

Syn. of Acid Magenta.

Rubramine—

(WM. NOETZEL AND CO.)—A derivative of Meldola's blue.—*Dark green powder,* soluble in water to a crimson solution ; in alcohol to a bright crimson solution ; in acetic acid to a dark crimson solution ; in strong sulphuric acid to a green solution, which on diluting with water turns to a grey blue solution. Hydrochloric acid turns the colour of the aqueous solution purplish blue, while caustic soda discharges the colour.—*Dyes* tannin mordanted cotton bluish crimson shades, turned blue by acids, maroon by caustic soda ; not fast to soaping ; fast to light.

S

Safranine—

(PERKIN.)—Usually a mixture of phenosafranine and tolusafranine

$$CH_3NH_2C_6H_3 \left\langle \begin{array}{c} N \\ | \\ N \end{array} \right\rangle C_6H_3CH_3$$

$$Cl \quad C_6H_4NH_2$$

obtained in several ways : 1st. Oxidation of Mauveine. 2nd. Heating aniline with glacial acetic acid and lead nitrate. 3rd. Heating of amido-azo-toluol with toluidine and nitric acid. 4th. Oxidation of equal molecules of *p*-toluylenediamine, aniline, and *o*-toluidine. (1863.) W. H. PERKIN.— *Red-brown powder,* easily soluble in water to a crimson solution ; in alcohol to a crimson solution, having a yellow-red fluorescence ; in strong sulphuric acid to a green solution ; on diluting with water this turns first blue then crimson. Hydrochloric acid added to the aqueous solution turns it blue-violet ; caustic soda gives a brown-violet precipitate. — *Dyes* cotton mordanted with tannin and tartar emetic, wool and silk in neutral baths fine bright shades of crimson ; those on silk have a slight fluorescence. Not fast to light, moderately so to washing and dilute acids. Also known as Aniline Rose, Pink.

Safranisol—

(KALLE.)—*P*-amido-*m*-methoxy-phenyl-*p*-amido-*m*-methoxy-phenazonium chloride

$$CH_3OC_6H_3 \left\langle \begin{array}{c} N \\ | \\ N \end{array} \right\rangle C_6H_3NH_2$$

$$Cl \quad C_6H_3NH_2OCH_3$$

obtained by oxidation of one molecule of *p*-phenylene diamine with two molecules of

ortho - anisidine. (1882.) NIETZKI.—Not now in use.

Safrosin—

(SOCY. CHEM. IND., BASLE.) — Syn. of Eosin BN.

St. Denis Red—

(POIRRIER.)—

Tetrazo-azoxy-ortho-toluidine $\left\langle \begin{array}{l} \alpha\text{-naphthol-}\alpha\text{-sodium} \\ \text{sulphonate} \\ \alpha\text{-naphthol-}\alpha\text{-sodium} \\ \text{sulphonate} \end{array} \right.$

$$O \left\langle \begin{array}{c} N-C_6H_3 \left\langle \begin{array}{c} CH_3 \\ N : NC_{10}H_5OHSO_3Na \end{array} \right. \\ N-C_6H_3 \left\langle \begin{array}{c} N : NC_{10}H_5OHSO_3Na \\ CH_3 \end{array} \right. \end{array} \right.$$

(1889.) *English Patent.*—*Dark red powder,* slightly soluble in water to a turbid solution ; in strong sulphuric acid to a deep scarlet solution ; on diluting with water a dark red precipitate is obtained. Hydrochloric acid completely precipitates the colour as a dark red precipitate ; caustic soda also gives a dark red precipitate.—*Dyes* cotton from an alkaline bath of salt dull scarlet, fast to acids, alkalies, and soaping.

Salicyl Orange—

(SCHERING.)—Sodium salt of monobromdinitro salicylic acid

$$C_6HCOONaONaBr(NO_2)_2$$

(1880.)—*Orange-yellow powder,* soluble in water to orange-yellow solution ; acids precipitate the free acid.—*Dyes* wool and silk orange shades. Not now in use.

Salicyl Yellow A—

(SCHERING.)—The monobromnitro salicylic acid

$$C_6H_2COOHOHBrNO_2$$

obtained by treating monobrom-salicylic acid with nitric acid. (1880.)—*Yellowish white powder,* very slightly soluble in water; soluble in alkalies.—*Dyes* wool and silk yellows, which are not very fast. Not now in use.

Salicyl Yellow B—

(SCHERING.)—The sodium salt of mono-bromnitro salicylic acid

$$C_6H_2COONaONaBrNO_2$$

treating the acid with soda. (See SALICYL YELLOW A). (1880.)—*Yellow powder,* soluble in water to a yellow solution ; acids precipitate the free acid.—*Dyes*

M

wool and silk in acid baths yellows, which are not fast. Not now in use.

Salmon Red—

(BADISCHE.)—Sodium salt of carbonyl-*p*-amido-benzene-azo naphthionic acid

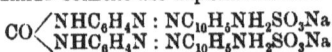

$$CO\begin{cases} NHC_6H_4N : NC_{10}H_5NH_2SO_3Na \\ NHC_6H_4N : NC_{10}H_5NH_2SO_3Na \end{cases}$$

obtained by treating *p*-amido-benzene-azo-naphthionic acid with phosgene gas. (1889.)—*English Patent* 14,222, 1889.—*Reddish brown powder*, slightly soluble in cold, easily soluble in hot water to a brownish red solution; in strong sulphuric acid to a purple solution; on diluting with water a reddish brown precipitate is obtained. Hydrochloric acid added to the aqueous solution gives a reddish brown precipitate; caustic soda an orange precipitate.—*Dyes* cotton from a neutral soap bath salmon to red-brown shades; acids turn it bluish green. Fast to alkalies and washing.

Scarlet GR—

(ACTIENGESELLSCHAFT.)—Sodium salt of xylene-azo-*b*-naphthol monosulphonic acid

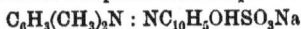

$$C_6H_3(CH_3)_2N : NC_{10}H_5OHSO_3Na$$

obtained by treating diazoxylene with *b*-naphthol monosulphonic acid S. (1879.)—*Scarlet-red powder*, soluble in water with a reddish yellow colour; in strong sulphuric acid with a cherry-red colour; on diluting with water a brown-red precipitate of the free sulphonic acid is obtained, and the same precipitate is formed on adding hydrochloric acid to the aqueous solution; caustic soda does not alter the colour. — *Dyes* wool in acid baths a yellowish scarlet.

Scarlet G—

(BADISCHE.)—Sodium salt of xylene-azo-*b*-naphthol disulphonic acid

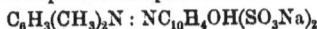

$$C_6H_3(CH_3)_2N : NC_{10}H_4OH(SO_3Na)_2$$

obtained by treating diazoxylene with *b*-naphthol disulphonic acid R. (1878.)—*Brown-red powder*, easily soluble in water; in strong sulphuric acid to a cherry-red solution; on diluting with water a reddish yellow solution is obtained; the aqueous solution is not altered on adding acids or caustic soda.—*Dyes* wool in acid baths a bright scarlet.

Scarlet J, JJ, V—

(MONNET.)—Syn. of Eosin BN.

Scarlet 3J—

Syn. of Ponceau 3G.

Scarlet R—

(BAYER.)—Syn. of Scarlet GR.

Silk Blues—

Are generally the sodium salts of the disulpho acid of Rosaniline Blues.

Smaragd Green—

(BAYER.)—Syn. of Brilliant Green.

Soap Yellow—

(POIRRIER.)—Sodium salt of *m*-amido-benzoic acid-azo-diphenylamine

$$C_6H_4COONaN : NC_6H_4NHC_6H_5$$

obtained by diazotising *m*-amido-benzoic acid and combining with diphenylamine. (1884.) (ROSENSTIEHL.)—*English Patent* 4,621, 1883.—*Brown powder*, very slightly soluble in water; soluble in strong sulphuric acid to a violet solution, which turns a bluish red on diluting with water; addition of hydrochloric acid to the aqueous solution turns it red-violet; caustic soda produces no change.—*Dyes* wool in a soap bath shades of orange, fast to soap.

Solferino—

Syn. of Magenta.

Solid Green—

(CASSELLA.)—Syn. of Malachite Green.

Solid Green—

(BADISCHE.)—See Resorcin Green.

Solid Green J—

Syn. of Brilliant Green.

Solid Yellow—

Syn. of Acid Yellow (2).

Soluble Aurine—

Syn. of Coralline.

Soluble Blue—

Syn. of Alkali and Water Blues.

Soluble Blue—

(SIMPSON, MAULE AND NICHOLSON.)—Sodium salt of triphenyl-*p*-rosaniline trisulphonic acid

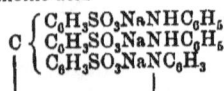

$$C\begin{cases} C_6H_3SO_3NaNHC_6H_5 \\ C_6H_3SO_3NaNHC_6H_5 \\ C_6H_3SO_3NaNC_6H_3 \end{cases}$$

acting with aniline upon pararosaniline

and sulphonating the blue produced. (1862.) E. C. NICHOLSON. *English Patent* June 24th, 1862.—*Dark blue powder,* soluble in water to a blue solution ; in strong sulphuric acid to a red-brown solution ; on diluting with water it regains its blue colour ; acids have no action on the aqueous solution ; caustic soda turns it red-brown.—*Dyes* wool and silks as alkali blues ; cotton requires a mordant of tannin and tartar emetic. Fine blue shades are obtained, fast to washing and acids ; moderately fast to light. Known also as Methyl Blue (CASSELLA), Methyl Cotton Blue, Brilliant Cotton Blue, Methyl Water Blue.

Soudan I—

(ACTIENGESELLSCHAFT.) — Benzeno-azo-*b*-naphthol

$$C_6H_5N:NC_{10}H_6OH$$

obtained by combining diazobenzene with *b*-naphthol, (1884.) LIEBERMANN.—*Red powder,* insoluble in water ; in strong sulphuric acid to a magenta-red solution; on diluting with water an orange-yellow precipitate is obtained. Soluble in alcohol. Used for making varnishes, lacquers.

Soudan II—

(ACTIENGESELLSCHAFT.)—Xylene-azo-*B*-naphthol

$$C_6H_3(CH_3)_2N:NC_{10}H_6OH$$

obtained by treating diazoxylene with *b*-naphthol. (1883.)—*Brown-red powder,* insoluble in water, soluble in strong sulphuric acid to a magenta-coloured solution; on diluting with water a pale yellow precipitate is obtained ; soluble in alcohol and fats with an orange colour, and is used for colouring varnishes and oils. Also known as Red B (BADISCHE).

Soudan G—

(ACTIENGESELLSCHAFT.) — *M*-dioxy-azo-benzene

$$C_6H_5N:NC_6H_3(OH)_2$$

obtained by diazotising aniline and combining with resorcin. (1875.)—*Brown powder,* partially soluble in water with a yellow colour. Hydrochloric acid produces a pale-brown precipitate ; caustic soda turns the colour of the aqueous solution brown ; soluble in strong sulphuric acid to a brownish yellow solution; on diluting with water a precipitate falls.

Soudan Brown—

(ACTIENGESELLSCHAFT.) — *A*-naphthyl-amino-azo-*a*-naphthol

$$C_{10}H_7N:NC_{10}H_6OH$$

(1878.) — *Brown powder,* insoluble in water ; soluble in alcohol and oils ; in strong sulphuric acid to a blue solution ; on addition of water it is thrown down as a brown precipitate ; slightly soluble in alkaline solutions with a wine-red colour. Used for colouring fats, etc. Also known as Pigment Brown (BADISCHE).

Spiller Purple—

(BROOKE, SIMPSON AND SPILLER.) — *Yellow bronze lustrous crystals,* slightly soluble in cold water ; easily soluble in boiling water to a violet solution ; soluble in alcohol and acetic acid to violet solutions ; in strong sulphuric acid to reddish brown solution, which on diluting with water gives a brownish yellow precipitate. Hydrochloric acid added in small quantity to the aqueous solution turns it an olive green ; caustic soda turns it a reddish violet.—*Dyes* tannin mordanted cotton, wool and silk from neutral baths fine purple-violet shades, tolerably fast to washing and light, not fast to acids and alkalies, which turn it an orange-red.

Spirit Blue—

(SIMPSON, MAULE AND NICHOLSON.)—The hydrochlorate, sulphate, or acetate of triphenyl rosaniline or pararosaniline

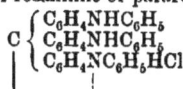

$$C \begin{cases} C_6H_4NHC_6H_5 \\ C_6H_4NHC_6H_5 \\ C_6H_4NC_6H_5HCl \end{cases}$$

obtained by acting with aniline upon a mixture of rosaniline and pararosaniline in the presence of acetic acid. (1860.) E. C. NICHOLSON. *English Patent* June 24th, 1862.—The hydrochlorate is a grey-green powder ; the sulphate and acetate are blue-violet powders ; insoluble in cold water ; very slightly in hot water ; soluble easily in alcohol ; soluble in strong sulphuric acid to a brownish yellow solution ; on diluting with water this gives a blue precipitate; acids added to the alcohol solution do not change it ; caustic soda turns it brown-red.—*Dyes* wool and silk in neutral baths fine shades of greenish blue, fast to acids and washing, and fairly fast to light. Known also as Aniline Blue (spirit soluble), Gentiana Blue 6B (ACTIENGESELLSCHAFT), Spirit Blue O (BADISCHE), Opal Blue, Night Blue, Fine Blue.

Spirit Eosin—

Syn. of Primrose.

Spirit Violet—

Syn. of Phenyl Violet.

Stanley Red—

(CLAYTON ANILINE CO.) — Ammonium salt of dehydrothioparatoluene - azo- B-naphthol

$$C_{14}H_{10}NS(SO_3NH_4)N:NC_{10}H_7OH$$

obtained by diazotising dehydro-*para*-thiotoluidine and combining with B-naphthol. (1889.) *English Patent* 18,901, 1889.— *Brownish red powder*, soluble in water to a scarlet solution; in strong sulphuric acid to a crimson solution; on diluting with water a scarlet precipitate is obtained. Hydrochloric acid added to the aqueous solution gives a scarlet precipitate. Caustic soda gives an orange precipitate. —*Dyes* silk from a slightly acid bath fine scarlet, turned maroon by acids and alkalies, not fast to soap and light.

Sterosine Grey—

(READ HOLLIDAY AND SONS.) — *Black liquid*, soluble in strong sulphuric acid to a maroon solution with copious evolution of hydrochloric acid gas ; on diluting with water a reddish black precipitate is obtained. Caustic soda added to the liquid causes the formation of dark purplish curdy masses.—*Dyes* unmordanted or mordanted cotton grey to grey-black shades very fast to acids, washing and light.

Sulfon Azurine D—

(BAYER.)—

Tetrazodiphenyl ⟋phenyl naphthylamine
  sulphone ⟍
  Sodium
  disulphonate ⟍ phenyl naphthylamine

$$SO_3NaC_6H_2N:NC_{10}H_6NHC_6H_5$$
$$\qquad\qquad \searrow SO_2$$
$$SO_3NaC_6H_2N:NC_{10}H_6NHC_6H_5$$

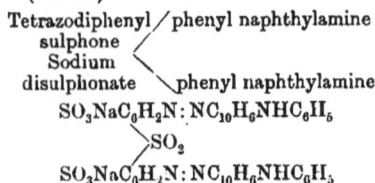

(1889.) *English Patent.—Violet powder*, slightly soluble in cold, easily in hot water, to a dark greenish blue solution; soluble in strong sulphuric acid to a deep purple solution which, on diluting with water, dyes cotton from a neutral soap bath greenish blues ; acids redden the shade; caustic soda darkens it. Fairly fast to washing, not to light.

Sulphanil Yellow—

(BAYER.)—

Tetrazodiphenyl ⟋sodium sulphanilate
  ⟍sodium sulphanilate

(1884.) *Brown-yellow powder*, soluble in water to a greenish yellow, not altered by acids or alkalies in the cold ; on warming with acids the dyestuff is decomposed,

nitrogen being evolved.—*Dyes* cotton from a soap bath greenish yellow, but the dyeings are weak and not fast. Not now in use.

Sun Yellow—

(GEIGY.)—Syn. of Curcumine S.

Sun Yellow—

Syn. of Heliochrysin.

T

Tartrazin—

(SOCY. CHEM. IND., BASLE, BADISCHE.)— Sodium salt of disulpho-diphenylizin-dioxytartaric acid

$$COOHC:NNHC_6H_4SO_3Na$$
$$\quad|$$
$$COOHC:NNHC_6H_4SO_3Na$$

obtained by acting with phenyl-hydrazin monosulphonic acid on dioxytartaric acid. (1885.)—*Fine orange-yellow powder*, easily soluble in water to a gold-yellow solution ; in strong sulphuric acid to a yellow solution, which does not change on being diluted with water ; acids added to the aqueous solution do not change it ; caustic soda turns it redder.—*Dyes* wool in acid baths fine chrome yellows, very fast to acids, light and washing. *Dyes* silk from an acid bath fine yellows, fast to light, etc.

Terra-cotta F—

(GEIGY.) — An azo-colouring matter, prepared from metaphenylene diamine-diazo-naphthionic acid and diazotised polychromine. (1890.) *English Patent* No. 1,688, 1890.—*Dark reddish brown powder*, soluble in water to a red solution; in alcohol to a dark amber solution ; only slightly soluble in acetic acid ; soluble in strong sulphuric acid to a red solution ; on dilution with water an orange-yellow precipitate falls down. Hydrochloric acid added to the aqueous solution throws down a brownish red precipitate ; caustic soda throws down a dull red precipitate.—*Dyes* unmordanted cotton from a boiling salt bath terra-cotta shades of red, turned violet-brown by acids and alkalies ; fast to souping, not fast to light on cotton ; rather faster on wool.

Thiazole Yellow—

(BAYER.) — Sodium amido-azo-dehydro-thio-toluidine sulphonate

$$CH_3C_6H_3 \underset{S}{\overset{N}{\diagdown}} CC_6H_3SO_3NaN : NNHC_6$$

$$H_3SO_3NaC \underset{S}{\overset{N}{\diagup}} C_6H_3CH_3$$

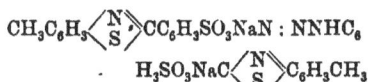

(1889.)—*Orange powder*, soluble in water to an orange yellow solution; in alcohol to a greenish yellow solution; in strong sulphuric acid to a brownish yellow solution; on diluting with water a dark orange precipitate is obtained. Hydrochloric acid added to the aqueous solution gives a dark orange precipitate; caustic soda a scarlet precipitate.—*Dyes* cotton from a salt bath fine greenish shades of yellow, turned orange by acids, red by alkalies. Not fast to washing.

Thiocarmine R—

(CASSELLA.) — Obtained from ethyl-benzyl-para-phenylene-diamine sulphonic acid by submitting it to Lauth's reaction

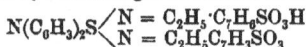

$$N(C_6H_3)_2S \underset{N\, =\, C_2H_5C_7H_3SO_3}{\overset{N\, =\, C_2H_5 \cdot C_7H_6SO_3H}{\diagup}}$$

(1890.) *British Patent.* — *Blue liquid, drying up to a mass with a bronzy colour*, soluble in water to a greenish blue solution; in strong sulphuric acid to a green solution; on diluting with water the blue colour is restored. Hydrochloric acid added to the aqueous solution does not alter it; caustic soda brightens the colour a little.—*Dyes* wool and silk from acid baths indigo carmine blue shades, not fast to light; fast to acids and weak alkalies; soaping slightly affects it.

Thioflavine S—

(CASSELLA AND CO.)—Sodium salt dimethyl-dehydro-thio-toluidine sulphonic acid. (1889.) *English Patent No. 14,884, 1888.* —*Brownish yellow powder*, slightly soluble in cold, easily in hot water, to a brownish yellow solution. Hydrochloric acid added to the solution throws down a red precipitate; caustic soda an orange-yellow precipitate.—*Dyes* cotton from a soap bath chrome yellows of a pure tone; silk and wool are dyed in a neutral bath; acids decolourise the fibre; fast to light, washing and alkalies.

Thioflavine T—

(CASSELLA AND CO.)—Trimethyl-dehydro-thio-toluidine chloride

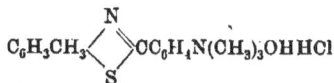

$$C_6H_3CH_3 \underset{S}{\overset{N}{\diagup}} CC_6H_4N(CH_3)_3OH\,HCl$$

(1889.) *English Patent 14,884, 1888. Orange-yellow powder*, soluble in water, alcohol and acetic acid to a yellow solution;

in sulphuric acid to a greenish solution; on diluting with water the original colour is restored. Hydrochloric acid has no action, caustic soda gives an orange-yellow precipitate. — *Dyes* cotton mordanted with tannin and tartar emetic fine greenish yellow; silk in a soap bath; the shades have a fine greenish fluorescence; wool from faintly acid baths fine light yellows, fast to dilute acids; strong acids decolourise them; caustic soda turns brown; fast to washing and light.

Thiorubin—

(DAHL.)—Sodium salt of thio-*p*-toluene-azo-*b*-naphthol disulphonic acid

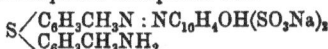

$$S \underset{C_6H_3CH_3NH_2}{\overset{C_6H_3CH_3N : NC_{10}H_4OH(SO_3Na)_2}{\diagup}}$$

obtained by diazotising thio-*p*-toluidine and combining with *b*-naphthol-disulphonic acid R. (1885.) — *Red-brown powder*, soluble in water to a magenta-red solution; in strong sulphuric acid to a crimson solution; on diluting with water a yellow-brown precipitate forms. On adding hydrochloric acid to the aqueous solution a yellow-brown precipitate is obtained; on adding caustic soda the colour becomes darker.—*Dyes* wool in acid baths a red.

Titan Brown R—

(READ HOLLIDAY AND SONS.)—A derivative of thio-para-toluidine sulphonic acid. (1890.) *English Patent No. 1,811, 1890.* —*Purple-brown powder*, soluble in boiling water and in alcohol to dull scarlet solutions; insoluble in acetic acid; soluble in strong sulphuric acid to a deep red solution; on diluting with water a pale orange turbid solution is obtained. Hydrochloric acid added to the aqueous solution gives an orange precipitate, while caustic soda produces no change.—*Dyes* unmordanted cotton from a salt bath terra-cotta shades of red, turned red-brown by strong acids, red by alkalies; fast to soaping, not fast to light.

Titan Brown Y—

(READ HOLLIDAY AND SONS.)—A derivative of thio-para-toluidine sulphonic acid. (1890.) *English Patent No. 1,811, 1890. — Purple-brown powder*, easily soluble in boiling water to an orange-red solution; in alcohol to a reddish amber; insoluble in acetic acid; strong sulphuric acid forms a crimson solution, which, on diluting with water, turns to a turbid orange solution. Hydrochloric acid added to the aqueous solution throws down an orange precipitate, as does also caustic soda.—*Dyes* cotton and wool from salt baths reddish browns, resembling Bismarck Brown ; turned brownish red by acids, reddened by alkalies ; fast to soaping ; not fast to light.

Titan Orange—

(READ HOLLIDAY AND SONS.)—A resorcin combination of thio-para-toluidine sulphonic acid. (1890.) *English Patent* No. 1,811, 1890.—*Brown crystalline powder*, soluble in water to an orange solution ; in alcohol to an amber-coloured solution ; insoluble in acetic acid ; soluble in strong sulphuric acid to a deep scarlet solution ; on dilution with water an orange-red turbid solution is obtained. Hydrochloric acid throws down an orange-red precipitate from the aqueous solution ; caustic soda turns the colour scarlet.—*Dyes* unmordanted cotton from salt baths orange shades, turned red by acids, crimson by alkalies ; fast to soaping ; not fast to light.

Titan Pink—

(READ HOLLIDAY AND SONS.)—A derivative of thio-para-toluidine monosulphonic acid. (1890.) *English Patent* No. 1,811, 1890.—*Dark brownish red powder*, soluble in water to a crimson solution ; in strong sulphuric acid to a bright magenta solution ; on diluting with water a dark red precipitate is obtained. Hydrochloric acid added to the aqueous solution gives a dark red precipitate ; caustic soda a brownish red precipitate. The commercial product is nearly chemically pure. —*Dyes* cotton from a salt bath shades from a pink to a bluish red ; the pale shades are most useful ; the dark shades are rather dull ; fast to soaping, acids and alkalies, and fairly fast to light and air.

Titan Pink BB—

(READ HOLLIDAY AND SONS.)—A derivative of thio-para-toluidine sulphonic acid. (1890.) *English Patent* No. 1,811, 1890. —*Brick-red powder*, soluble in water to a dull red solution ; in acetic acid to a bluish red solution ; in strong sulphuric acid to a crimson solution ; on diluting with water a brownish red precipitate is obtained. Hydrochloric acid added to the aqueous solution turns it a little yellower ; caustic soda gives a crimson precipitate.—*Dyes* wool from a salt bath crimson red, fast to acids ; turned a little browner by alkalies ; not fast to soaping.—*Dyes* cotton from a salt bath bluish pinks, brightened by soaping ; fast to acids and alkalies ; fairly fast to light.

Titan Red—

(READ HOLLIDAY AND SONS.)—A derivative of thio-para-toluidine sulphonic acid. (1890.) *English Patent* No. 1,811, 1890. *Dark brownish red powder*, soluble in water to a red solution, in strong sulphuric acid to a crimson solution ; on diluting with water a scarlet precipitate forms. Hydrochloric acid added to the aqueous solution gives a scarlet precipitate ; caustic soda has no action.—*Dyes* cotton dull magenta-red ; wool bright bluish red from a salt bath ; fast to acids and alkalies ; moderately fast to light and soaping.

Titan Scarlet—

(READ, HOLLIDAY AND SONS.)—A derivative of thio-para-toluidine sulphonic acid. (1890.) *English Patent* No. 1,811, 1890. *Brick-red powder*, soluble in water to a scarlet solution ; in strong sulphuric acid to a crimson solution ; on diluting with water this turns orange. Hydrochloric acid added to the aqueous solution makes it more turbid ; caustic soda turns it a dull scarlet.—*Dyes* cotton from a salt bath a dull scarlet ; wool from a faintly acid salt bath bright scarlet ; fast to acids and alkalies ; not fast to light ; moderately fast to soaping.

Titan Scarlet B—

(READ HOLLIDAY AND SONS.)—A derivative of thio-para-toluidine sulphonic acid. (1890.) *English Patent* No. 1,811, 1890. —*Brick-red powder*, slightly soluble in cold, soluble in hot water to a turbid red solution ; in strong sulphuric acid to a violet solution ; on diluting with water a reddish brown precipitate is obtained. Hydrochloric acid added to the aqueous solution gives a reddish brown precipitate ; caustic soda has no action.—*Dyes* wool from a salt bath bright bluish scarlet, fast to dilute acids ; turned browner by strong acids and alkalies ; not fast to soaping. On cotton the shades are poor.

Titan Scarlet C—

(READ HOLLIDAY AND SONS.)—A derivative of thio-para-toluidine monosulphonic acid. (1891.) *English Patent* No. 1,811, 1890.—*Dark red crystalline powder*, soluble in water to a red solution ; in alcohol to a pale red solution ; in acetic acid to a red solution ; in strong sulphuric acid to a crimson solution, which on diluting with water gives an orange - coloured turbid solution. Hydrochloric acid added to the aqueous solution throws down a red precipitate, while caustic soda throws down a crimson flocculent precipitate.— *Dyes* cotton, wool, and silk from salt baths reds, not bright ; fast to acids and soaping, but not to light.

Titan Yellow—

(READ HOLLIDAY AND SONS.)—A combination of thio-para-toluidine sulphonic acid with a phenol. (1890.) *English Patent* No. 1,811, 1890.—*Orange-brown powder*, soluble in water to an amber-coloured solution; in strong sulphuric acid to a scarlet solution ; on diluting with water a brown precipitate is obtained. Hydrochloric acid added to the aqueous solution gives a brown flocculent precipitate. Caustic soda turns the colour of the solution orange.—*Dyes* cotton and wool from salt baths fine bright orange-yellow, reddened by acids and alkalies, moderately fast to soaping, the shade is reddened slightly; fast to light.

Titan Yellow Y—

(READ HOLLIDAY AND SONS.)—A combination of thio-para-toluidine sulphonic acid with a phenol. (1890.) *English Patent* No. 1,811, 1890.—*Brown powder*, soluble in water to a brownish yellow ; in alcohol to a reddish orange ; in acetic acid to a pale yellow solution ; in strong sulphuric acid to an orange scarlet solution ; on diluting with water it turns yellow, and gives a brown precipitate. Hydrochloric acid added to the aqueous solution gives a reddish yellow precipitate ; caustic soda gives a red precipitate.—*Dyes* wool and cotton a chrome yellow from a salt bath, turned orange by acids, scarlet by alkalies ; fast to soaping, which does not alter the shade ; fairly fast to light.

Tolidine—

$$C_6H_3CH_3NH_2$$
$$|$$
$$C_6H_3CH_3NH_2$$

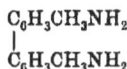

is a diad base prepared from ortho-hydrazo toluene, and can be got in *colourless crystals*, melting at 128deg. C. It is used in the manufacture of some of the so-called direct cotton colours. See BENZIDINE COLOURING MATTERS.

Toluene—

Or methyl-benzene

$$C_7H_8 \text{ or } C_6H_5CH_3.$$

An aromatic hydrocarbon found in coal tar, the second member of the benzene series of hydrocarbons.—*A limpid liquid*,

having a peculiar odour, resembling that of benzene. Sp. gr., 0·882. Boiling point, 111deg. C.; other properties, solubility, etc., resemble those of benzene. When treated with nitric acid it is converted into nitrotoluene

$$C_6H_4NO_2CH_3$$

which, being a di-derivative of benzene forms three isomers, *ortho, para* and *meta*. Reduced by nascent hydrogen, these yield *ortho, para* and *meta* toluidine respectively.

Toluidine—

Amidotoluene

$$C_6H_4{}^{NH_2}_{CH_3}$$

This substance, being a di-derivative of benzene, exists in three isomeric modifications ortho, meta, and para. It is obtained by the reduction with iron filings and hydrochloric acid of the nitro-toluenes. Commercial toluidine consists principally of ortho and para toluidine ; there is but little of the meta product.

Ortho-toluidine is a colourless liquid ; boiling at 197—197·5deg. C., sp. gr. is 0·9978 ; solidifies below—20deg. C. ; on exposure to air it becomes more or less brown; it is slightly soluble in water; freely in alcohol and ether ; dissolves in acids forming the toluidine salts. Ortho-toluidine is present in most varieties of commercial aniline oil ; it is used for preparing several colouring matters ; when heated with sulphur it yields thio-toluidines, noticeable on account of their affinity for the cotton fibre. See PRIMULINE.

Meta-toluidine is also a colourless oil ; boiling at 197deg. C. ; having a sp. gr. of 0·998, it solidifies at—13deg. C.

Para-toluidine forms colourless tabular crystals ; melting at 45deg. C.; boiling at 198deg. C. ; sparingly soluble in water ; readily in alcohol and ether. It forms salts with acids although it is only a weak base. It finds a use in the manufacture of magenta.

The two toluidines, ortho and para, are separated partly by a process of freezing, whereby the para crystallises out to a large extent, and is easily separated out ; the liquid ortho still contains some para—this is separated out by partially acidifying with oxalic acid, and then distilling with steam ; the para-toluidine first combines with the acid, and is then not distillable with the steam, while the ortho-toluidine is driven over.

There are other methods of separating the two, converting them into acetotoluides ; the ortho product is soluble in water, the para is not and separates out ; adding hydrochloric acid and phosphate of soda ; the ortho-toluidine remains un-altered ; the para-toluidine and any aniline that is present crystallise out as phosphates. *Aniline oil for red* consists of about equal molecular proportions of ortho-para-toluidine and aniline. *Aniline oil for safranine* consists chiefly of aniline and ortho-toluidine, although a small quantity of para-toluidine is present. Benzylamine $C_6H_5CH_2NH_2$ is metameric with Toluidine.

Toluylene Blue—

A blue of the Indamine class, obtained by treating meta-toluylene-diamine with nitroso-dimethyl aniline in aqueous solution. (1879.) WITT. *Jour. Chem. Socy.*, 1879, Vol. xxxv., p. 356.—*Black-brown needles*, dissolving in water and alcohol to a green-blue solution ; in strong sulphuric acid to a yellow solution. This blue has not come into use as it is unstable.

Toluylene Blue B—

(OEHLER.)—1891.—*Dark violet powder*, soluble in water to a violet solution ; in alcohol to a blue solution ; in acetic acid it dissolves with a blue colour ; strong sulphuric acid forms a dark blue solution, which on dilution with water forms a turbid violet solution. Hydrochloric acid added to the aqueous solution has no action, while caustic soda decolourises it, throwing down a white precipitate of the free base.—*Dyes* tannin mordanted cotton fine deep blue shades, slightly reddened by acids ; fast to alkalies, soap and light.

Toluylene Blue R—

(OEHLER.) — A water soluble basic induline. (1891.)—*Violet - black powder*, soluble in water to a blue solution ; in alcohol and acetic acid to deep blue solutions ; in strong sulphuric acid to a deep reddish blue solution, which on diluting with water gives a turbid purple solution. Hydrochloric acid has no action on the aqueous solution, while caustic soda discharges the colour and throws down a white precipitate of the colour base.—*Dyes* tannin mordanted cotton deep reddish shades of blue, turned bluer by acids ; fast to alkalies, soap and light.

Toluylene Brown B—

(OEHLER.)—A combination of sulphonic acid of Bismarck Brown with diazotised sulphanilic acid or an analogue. The sulphonic acid of Bismarck Brown is made from toluylene diamine sulphonic acid. (1891.)—*Reddish brown powder*, slightly soluble in cold water, readily soluble in boiling water to a reddish brown solution ; soluble in alcohol to a pale wine-red solution ; in strong sulphuric acid to a dark purple solution, which on diluting with water becomes very pale. Hydrochloric acid added to the aqueous solution throws down a reddish brown precipitate ; caustic soda also throws down a reddish brown precipitate.—*Dyes* cotton from a salt and soap bath giving dark shades of brown ; fast to soaping ; acids turn it a dark brown while alkalies change it to a reddish brown ; not fast to light.

Toluylene Brown M—

(OEHLER.)—Has a similar composition to Toluylene Brown B. (1891.)—*Pale reddish brown powder*, slightly soluble in cold water ; soluble on boiling to a pale brownish red solution ; soluble in alcohol to a pale brownish red solution ; strong sulphuric acid forms a dark purple solution, which on diluting with water turns pale brown. Hydrochloric acid has no action on the aqueous solution, while caustic soda throws down a reddish precipitate.—*Dyes* unmordanted cotton from a boiling soap bath reddish brown shades, fast to soaping, turned dark brown by strong acids, reddish brown by alkalies ; not fast to light.

Toluylene Brown R—

(OEHLER.)—Similar in composition to Toluylene Brown B. (1891.) — *Dark brown powder*, slightly soluble in cold, easily soluble in hot water to a pale reddish brown solution ; in alcohol it dissolves with a pale brownish red colour ; strong sulphuric acid dissolves it, forming a blackish purple solution, which on diluting with water turns to a pale brown. Hydrochloric acid turns the colour of the aqueous solution a little yellower ; caustic soda throws down a pale brownish red precipitate.—*Dyes* unmordanted cotton from a boiling bath of soap and salt orange-browns, turned dark brown by acids, redder by alkalies ; fast to soaping, not fast to light.

Toluylene Orange G—

(OEHLER.)—1888. *English Patent* 4,492, 1887.—*Dark orange powder*, very slightly soluble in cold water, soluble on boiling to an orange solution ; slightly soluble in alcohol to an orange solution ; insoluble in acetic acid ; soluble in strong sulphuric acid to a fine purple solution ; on diluting with water a brown curdy precipitate falls down. Hydrochloric acid added to the aqueous solution throws down a curdy reddish brown precipitate, while caustic soda gives a bright orange, somewhat turbid, solution. —*Dyes* cotton from a neutral soap bath, wool and silk from salt baths ; reddish shades of yellow are got, fast to soap, turned brown by acids, red by alkalies.

Toluylene Red—
Syn. of Neutral Red.

Tropæolin D—
Syn. of Orange III.

Tropæolin O (*original name*)—
Syn. of Resorcin Yellow.

Tropæolin OO (*original name*)—
Syn. of Acid Yellow I.

Tropæolin OOO No. 1 (*original name*)—
Syn. of Orange I.

Tropæolin OOO No. 2 (*original name*)—
Syn. of Orange II.

Tropæolin OOOO—
The sodium salt of benzene-azo-a-naphthol monosulphonic acid

$$C_6H_5N:NC_{10}H_5OHSO_3Na$$

obtained by combining diazobenzene with a-naphthol monosulphonic acid N.W. (1877.) WITT. *Chem. Socy. Jour.* 1879, p.184.—*Red-brown powder*, soluble in water to a reddish yellow solution ; in strong sulphuric acid it dissolves with a magenta colour ; on adding water no colour change is produced. Hydrochloric acid added to the aqueous solution produces a yellowish red precipitate. Caustic soda changes the colour of the aqueous solution to orange-yellow. — *Dyes* wool and silk in acid baths a dull red. Not now in use. Known also as Azo Coccin G.

Tropæolin R—
Syn. of Resorcin Yellow.

Tropæolin Y—
Sodium salt of sulphanilic acid azophenol

$$C_6H_4SO_3NaN:NC_6H_4OH$$

obtained by diazotising sulphanilic acid and combining with phenol. (1873.) TSCHIRVINSKY, 1875. — GRIESS, WITT.

Chem. Socy. Jour., 1879, p. 184.— *Brownish yellow powder*, soluble in water to a reddish yellow solution, and in strong sulphuric acid to a yellow solution ; on diluting with water a brownish yellow solution is obtained ; on adding hydrochloric acid to the aqueous solution the colour becomes darker ; caustic soda also turns the colour darker.—*Dyes* wool in acid baths a yellow, not fast. Not now in use.

Turmerine—
(BROOKE, SIMPSON AND SPILLER.)— 1890.—*Brownish yellow powder*, soluble in water to a brownish yellow solution ; in strong sulphuric acid to an amber-coloured solution, which, on diluting with water, turns pale brownish yellow. Hydrochloric acid added to the aqueous solution gives an orange precipitate ; caustic soda gives an orange-coloured, rather turbid, solution.—*Dyes* cotton and wool from salt bath chrome yellow, turned orange by acids, scarlet by alkalies ; fast to soaping and light.

Tyralin—
An old name for Mauvein.

U

Uranine—
Sodium salt of fluorescein tetraoxy-phthalophenone-anhydride

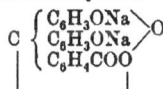

$$C \left\{ \begin{matrix} C_6H_3ONa \\ C_6H_3ONa \\ C_9H_4COO \end{matrix} \right\rangle O$$

obtained by heating resorcin with phthalic anhydride, either alone or in presence of chloride of zinc. (1877.)—*Yellow-brown powder*, easily soluble in water to a yellow solution, which has a most intense yellow green fluorescence, easily soluble in alcohol ; soluble in strong sulphuric acid to a yellow solution, with a feeble fluorescence ; on diluting with water there is little change. Hydrochloric acid gives in the aqueous solution a yellow flocculent precipitate ; caustic soda turns the colour of the aqueous solution darker, and the alkaline solution has a dark green fluorescence.—*Dyes* wool in acid baths yellow ; silk in acid baths yellow, with a fluorescence.

Uranine—
Under this name methyl fluorescoine is sent out.

Usebe Green—
Syn. of Aldehyde Green.

N

V

Vesuvin—

Syn of Bismarck Brown.

Victoria Black

(BAYER.) — 1891. — *Greyish black powder*, soluble in water to a dark purple solution ; in alcohol to a deep blue solution ; in acetic acid to deep blue solution ; in strong sulphuric acid to an olive-green solution, which on dilution with water gives a dark purple precipitate. Hydrochloric acid turns the colour of the aqueous solution a pale red ; on standing, a precipitate falls out. Caustic soda changes the colour of the solution to a violet-blue.—*Dyes* wool from acid baths deep navy blue to blue-black, fast to acids and alkalies ; not quite fast to strong soaping ; fast to light and air.

Victoria Black 5G—

(BAYER.) — 1891. — *Blackish brown powder*, slightly soluble in cold water to a turbid,reddish black solution ; insoluble in alcohol and acetic acid ; soluble in strong sulphuric acid to an olive-green solution ; on diluting with water a purplish black precipitate falls down. Hydrochloric acid has no action on the aqueous solution, while caustic soda throws down a dark blue precipitate.—*Dyes* wool from acid baths blue-grey to blue-black shades ; turns blue by acids and alkalies ; not fast to strong soaping ; fast to light.

Victoria Blue B—

(BADISCHE, SOCY. CHEM. IND., BASLE.) — Hydrochlorate of phenyl-tetramethyl triamido-a-naphthyl-diphenyl-carbinol

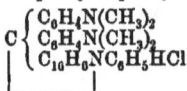

$$C \begin{cases} C_6H_4N(CH_3)_2 \\ C_6H_4N(CH_3)_2 \\ C_{10}H_6NC_6H_5HCl \end{cases}$$

obtained by acting with phenyl-a-naphthylamine upon tetramethyl-diamido-benzophenone chloride. (1883.) CARO AND KERN. — *English Patent* 5,003, 1884 ; 11,159, 1884.—*Bronzy lustrous crystals*, slightly soluble in cold water ; more freely in hot water ; soluble in strong sulphuric acid to a reddish yellow solution ; on diluting with water it turns first yellow, then finally blue ; soluble in alcohol to a pure blue solution. Hydrochloric acid added to the aqueous solution turns it first green then dark yellow brown ; caustic soda gives a red-brown precipitate. —*Dyes* wool in an acid bath ; silk in a broken soap or sulphuric acid bath ; cotton either mordanted with tannin and

tartar emetic or direct in an acetic acid bath. The shades given are a pure, brilliant blue, fairly fast to light, air and washing. Also known as Victoria Blue BS.

Victoria Blue BS—

Syn. of Victoria Blue B.

Victoria Blue 4R—

(BADISCHE, SOCY. CHEM. IND., BASLE.)— Hydrochlorate of phenyl-penta-methyl-triamido-a-naphthyl-diphenyl-carbinol

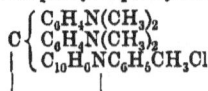

$$C \begin{cases} C_6H_4N(CH_3)_2 \\ C_6H_4N(CH_3)_2 \\ C_{10}H_6NC_6H_5CH_3Cl \end{cases}$$

(1883.) CARO AND KERN.—*English Patent* 5,003, 1884 ; 11,159, 1884. — *Bronzy lustrous powder, or in yellowish bronzy lumps*, soluble in warm water to a blue violet solution ; in strong sulphuric acid to a yellow-brown solution ; on diluting with water this turns first green, then blue. Hydrochloric acid changes the colour of the aqueous solution first green, then yellow-brown. Caustic soda gives a violet-brown precipitate.—*Dyes* wool, etc., the same way as Victoria Blue B, but gives rather more violet shades.

Victoria Green—

(BADISCHE.)—Syn. of Malachite Green.

Victoria Green 3B—

(BADISCHE.)—Hydrochlorate of tetra-methyl-diamido-triphenyl-carbinol sul-phonic acid

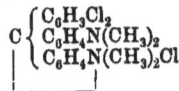

$$C \begin{cases} C_6H_3Cl_2 \\ C_6H_4N(CH_3)_2 \\ C_6H_4N(CH_3)_2Cl \end{cases}$$

obtained by acting with dichlor-benzalde-hyde upon dimethyl aniline and oxida-tion of the tetramethyl-diamido-dichlor-triphenylmethan so produced. An older process consisted in acting with chloro-benzo-trichloride upon dimethyl aniline. (1878.) ACTIENGESELLSCHAFT.(1883.) SOCY. CHEM. IND., BASLE.—*Green crystalline powder, with a bronzy metallic lustre*, very slightly soluble in cold water, easily in warm water, to a bluish green solution ; in strong sulphuric acid to a yellow solu-tion which, on diluting with water, turns first reddish yellow then yellow-green. Hydrochloric acid added to the aqueous solution turns it first yellow-green then yel-low. Caustic soda precipitates the base. Soluble in alcohol.—*Dyes* cotton mor-danted with tannin and tartar emetic bluish green ; wool and silk in neutral baths. The shades are fairly fast to light, air and washing. Also known as New Solid Green 3B.

Victoria Orange—

Syn. of Victoria Yellow.

Victoria Yellow—

The potassium salt of dinitro-para-cresol

$$C_6H_2CH_3(NO_2)_2KO$$

obtained by treating para-cresol or para-toluidine with nitric acid. (1869.)—*Red crystals*, dissolving in water with a yellow colour. Hydrochloric acid gives a white precipitate of dinitro-cresol. Dissolves in strong sulphuric acid with a yellow colour.—*Dyes* wool and silk yellows of a reddish tone in an acid bath. Not fast, especially to washing. Rarely used. Also known as English Yellow, Victoria Orange, Aniline Orange.

Violamine B—

(MEISTER LUCIUS.)—Is a sodium salt of a sulphonic acid of a basic dyestuff derived from phthalein. (1890.)—*Purple powder*, soluble in water to a dark crimson solution ; in strong sulphuric acid to a dull red solution, which on diluting with water turns crimson. Hydrochloric acid added to the aqueous solution turns it violet ; caustic soda turns it crimson.—*Dyes* wool and silk from acid baths brilliant reddish violet ; turns bluer by acids and redder by alkalies ; not fast to soaping.

Violamine 2R—

(MEISTER LUCIUS.)—1890.—*Dull purple-red powder*, soluble in water to a scarlet solution ; in alcohol to a bluish pink solution having an orange fluorescence ; in strong sulphuric acid to an amber-coloured solution, which on diluting with water turns bluish pink. Hydrochloric acid turns the colour of the aqueous solution yellower, caustic soda bluer.—*Dyes* wool and silk from acid baths light crimson shades very bright ; fast to acids and alkalies, not fast to soaping.

Violaniline—

The parent colour base of the Indulines.

Violet Black—

(BADISCHE.)

Phenylene tetrazo $\Big<$ begin{array}{l} α\text{-naphthol sodium monosulphonate N.W.} \\ α\text{-naphthylamine} \end{array}$

(C_6H_4)

obtained by combining acetanilide with

α-naphthol sodium monosulphonate N.W., displacing the acetyl group, diazotising the base so produced, and combining with α-naphthylamine. (1887.)—*Bronzy lustrous powder*, soluble in water to a brown-red solution ; in strong sulphuric acid to a blue solution ; on diluting with water a violet precipitate is obtained. Hydrochloric acid added to the aqueous solution gives a violet precipitate, while caustic soda gives a red-violet colour.— *Dyes* cotton from soap bath, and wool from a salt bath, violet-black shades, fast to light, acids and washing.

Violet 5B—

Syn. of Benzyl Violet.

Violet 6B—

Syn. of Crystal Violet.

Violet 5R—

Syn. of Hofman's Violet.

Violet RR—

Syn. of Hofman's Violet.

Viridine—

(BROOKE, SIMPSON AND SPILLER.)—Syn. of Alkali Green.

W

Water Blue—

(SIMPSON, MAULE AND NICHOLSON.)— Sodium, ammonium or calcium salt of triphenyl-rosaniline, or para-rosaniline trisulphonic acid

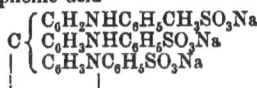

$$C\begin{cases} C_6H_2NHC_6H_5CH_3SO_3Na \\ C_6H_3NHC_6H_5SO_3Na \\ C_6H_3NC_6H_5SO_3Na \end{cases}$$

obtained by acting with strong sulphuric acid on Spirit Blue. (1862.) E. C. NICHOLSON. *English Patent* June 1, 1862. —*Blue powder or granules with a metallic lustre*, easily soluble in water to a blue solution ; in strong sulphuric acid to a reddish yellow solution ; on diluting with water a blue solution and a blue precipitate is obtained. Hydrochloric acid only gives a precipitate in the aqueous solution when the disulphonic acid is present. Caustic soda gives a brown-red solution. —*Dyes* wool and silk in acid baths ; cotton in a boiling bath with alum, or better mordanted with tannin and tartar emetic. Fine greenish blue shades are obtained ; fast to acids and washing ; moderately fast to light. Known also as Cotton Blue, China Blue, Navy Blue, Opal Blue, Soluble Blue.

Wool Black—

(ACTIENGESELLSCHAFT.)—Sodium salt of benzene sulphonic acid azo-benzene sulphonic azo-p-tolyl-b-naphthylamine

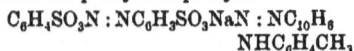

$$C_6H_4SO_3N : NC_6H_3SO_3NaN : NC_{10}H_6$$
$$NHC_6H_4CH_3$$

obtained by diazotising amido-azo-benzenedisulphonic acid and combining with p-tolyl-b-naphthylamine. (1885.) L. SCHAD. *English Patent* 9,754, 1886.—*Blue-black powder*, soluble in water to a dark violet-blue solution ; in strong sulphuric acid to a blue solution ; on diluting with water a brown precipitate falls ; by boiling this diluted acid solution decomposition occurs with the formation of amido-azo-benzenedisulphonic acid and toluene naphthazin. Addition of hydrochloric acid to the aqueous solution produces a red-violet precipitate ; of caustic soda a violet precipitate.—*Dyes* wool in acid baths a dark blackish blue, fast to washing and acids. ,Also known as Wool Black (BADISCHE).

Wool Black—

(BADISCHE.) — Syn. of Wool Black (ACTIENGESELLSCHAFT).

Wool Scarlet R—

(SCHOLLKOPF.)—Sodium salt of xylene-azo-a-naphthol disulphonic acid.

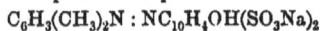

$$C_6H_3(CH_3)_2N : NC_{10}H_4OH(SO_3Na)_2$$

obtained by treating diazoxylene with a - naphthol disulphonic acid.　(1884.) *English Patent* 15,775, 1885.—*Brown-red powder*, soluble in water to a yellowish red solution ; in strong sulphuric acid to a cherry-red solution ; on diluting with water a red solution is obtained ; on adding hydrochloric acid the solution changes colour to a blue-red ; caustic soda changes the colour to a strong yellowish red.—*Dyes* wool and silk in acid baths a fine scarlet red, fast to washing, etc.

Wool Scarlets, O, OO, OOO, OOOO—

(READ HOLLIDAY AND SONS.)—A range of scarlets yielding bright shades, prepared from a new naphthylamine sulphonic acid. (1889.) *English Patent.*

X

Xylene—

Or Dimethylbenzene

$$C_8H_{10} \text{ or } C_6H_4(CH_3)_2.$$

An aromatic hydrocarbon, the third member of the benzene series found in coal tar. It exists in three isomeric modifications—*ortho, meta,* and *para.* ·

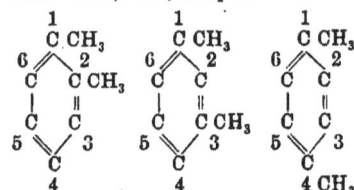

Ortho (1: 2)　Meta (1: 3)　Para (1: 4)

They are colourless volatile liquids, with an odour resembling benzene. Ortho-xylene boils at 141-142deg. C., meta-xylene at 140deg. C., and para-xylene at 136deg. C. All three are present in the commercial product obtained from coal tar ; of these the meta-xylene is the most abundant and technically the most important. It forms from 70 to 87 per cent. of ordinary commercial xylene ; the ortho-xylene being present in quantities varying from 2 to 15 per cent., and the para-xylene from 3 to 10 per cent. (LEVINSTEIN, *Jour. Socy. Chem. Ind.,* 1884, p. 79.) Each of these treated with nitric acid gives a nitro-xylene, and from these the ortho, para and meta xylidines can be obtained.

Xylidine—

Amido-xylene

$$C_6H_3(CH_3)_2NH_2.$$

Theoretically there are six isomeric xylidines possible, two from each variety of xylene. The commercial xylidine is a *dark coloured oily liquid*, boiling between 211deg. C. and 219deg. C., and having a sp. gr. of 0·978 ; it usually consists of a mixture in various proportions of five isomeric xylidines, one para xylidine, two ortho, and two meta xylidines. It is used in the manufacture of many of the azo-scarlets.

Xylidine Red—

Syn. of Ponceau 2R.

Xylidine Scarlet—

Syn. of Ponceau 2R.

Y

Yellow T—

(SOCY. CHEM. IND., BASLE.)—Syn. of Resorcin Yellow.

Yellow Coralline—

Syn. of Coralline.

ADDENDA.

Anthracene Yellow—

(BAYER.)—Double brominated dioxy-*b*-methyl coumarine. (1890.)—*Grey liquid*, the colouring matter in which is insoluble in water; soluble in alcohol to a yellow solution; insoluble in acetic acid; strong sulphuric acid dissolves it to an olive-brown solution; on dilution with water the original grey colour comes back. Hydrochloric acid has no action on the aqueous solution; caustic soda gives an orange-brown solution.—*Dyes* chrome or alumina mordanted wool olive-yellows, rather dull; darkened by acids; turned brownish yellow by caustic soda; fast to light and soaping. Can be used in calico and wool printing with chrome and alumina mordants, giving moderately fast shades.

Anthracite Black B—

(CASSELLA AND CO.) — 1891. *English Patent* No. 7,977, 1889.—*Brownish black crystalline powder*, soluble in water to a violet-black solution; in alcohol to a deep navy blue solution; in acetic acid to a dark crimson solution; in strong sulphuric acid to a blackish blue solution, from which, on dilution with water, a brownish olive precipitate falls down. Hydrochloric acid changes the colour of the aqueous solution to a navy blue; caustic soda throws down a dark violet-red precipitate.—*Dyes* wool and silk from acid baths grey to black shades of a bluish tone; fast to acids; turned bluer by alkalies; not fast to soaping, fast to light.

Azamine 4B—

(HEYWOOD CHEMICAL CO.)—1891. *English Patent* Nos. 12,767, 1889; 14,294, 1889; 3,095, 1890.—*Dark red powder*, not readily soluble in cold water; more soluble in hot water to a turbid scarlet solution; in alcohol it forms a turbid red solution; acetic acid turns the colour of the powder to brown; strong sulphuric acid dissolves it to a blue solution, from which on diluting with water a blackish blue precipitate falls down. Hydrochloric acid throws down the same precipitate from an aqueous solution; caustic soda throws down a scarlet precipitate.—*Dyes* unmordanted cotton, wool and silk from boiling soap and soda or salt baths scarlet shades resembling

Turkey-red; turned dark blue by acids; slightly yellower by caustic soda; not quite fast to soaping; not fast to light and air.

Azo-Acid Brown—

(BAYER.) — 1891.—*Pale yellow-brown powder*, slightly soluble in cold water, easily in hot water to a dark yellow-brown solution; soluble in alcohol to a turbid yellow-brown solution; in acetic acid to a turbid dark brown solution. Strong sulphuric acid dissolves it to a very deep violet-blue solution, from which a dark brown precipitate falls down on adding water. Hydrochloric acid throws down from the aqueous solution a dark brown precipitate. Caustic soda produces no change.—*Dyes* wool or silk from acid baths shades of yellow brown, similar to Bismarck Brown; acids turn them dark brown; alkalies slightly redden; soaping causes them to bleed a little.

Azo-Acid Violet 4R—

(BAYER.)—1891.—*Brown powder of a slight reddish hue*, soluble in water to a crimson solution; in alcohol and acetic acid to bluish crimson solution; in strong sulphuric acid to a deep crimson solution, from which, on diluting with water, a dirty red precipitate falls down. Hydrochloric acid added to the aqueous solution throws down a dull red precipitate; caustic soda changes the colour of the solution to scarlet.—*Dyes* wool and silk from acid baths dull crimson shades, turned slightly darker by acids; bright orange by alkalies; not fast to soaping.

Azo Mauve—

(OEHLER.)—1891. — *Brown crystalline powder*, slightly soluble in cold water; easily soluble in boiling water to a reddish violet solution; in alcohol to a deep reddish violet solution; in acetic acid to a pale reddish violet solution; in strong sulphuric acid to a deep blue solution from which, on diluting with water, a violet precipitate falls down. Hydrochloric acid added to the aqueous solution throws down a dark red violet precipitate; caustic soda throws down a brown violet precipitate.—*Dyes* cotton and wool from a boiling salt bath a violet of a reddish hue, not bright, not fast to soaping or light; darkened by acids and alkalies.

Benzo Brown BR—

(BAYER.)—1891.—*Dark brown powder*, insoluble in cold, easily soluble in hot water to a dark reddish brown solution ; in alcohol to a brownish scarlet solution ; insoluble in acetic acid ; soluble in strong sulphuric acid to a deep navy blue solution, which turns reddish brown on dilution with water. Hydrochloric acid added to the aqueous solution throws down a reddish brown precipitate ; caustic soda has no action on the aqueous solution.— *Dyes* cotton from a salt bath dull terracotta shades of red ; dyes wool from salt baths dark brownish reds, turned dark brown by acids, darker by caustic soda. Fairly fast to soaping ; not fast to light.

Benzo Brown 5R—

(BAYER.)—1891. —*Brick - red powder*, slightly soluble in cold, easily soluble in hot water to a brownish orange solution, in alcohol to a brownish amber solution ; insoluble in acetic acid ; soluble in strong sulphuric acid to a scarlet solution which turns reddish orange on diluting with water. Hydrochloric acid added to the aqueous solution throws down a brownish orange precipitate ; caustic soda also throws down same precipitate.— *Dyes* cotton and wool from salt baths dark reddish orange, turned blackish brown by acids, brownish red by caustic soda ; fairly fast to soaping, not fast to light.

Benzo Indigo Blue—

(BAYER.)—1891.—*Grey powder*, soluble in cold water to a dull blue solution ; in alcohol to a turbid reddish blue solution ; in acetic acid to a dark blue turbid solution ; in strong sulphuric acid to a dark greenish blue solution, from which, on diluting with water, a blue precipitate falls down. Hydrochloric acid added to the aqueous solution throws down a blue precipitate, while caustic soda changes it to a turbid reddish blue solution.—*Dyes* unmordanted cotton from a salt and soap bath bright indigo blue shades ; on wool it gives a violet shade of blue ; fast to soaping and acids ; turned brown by alkalies ; not fast to light.

Benzo Olive—

(BAYER.)—1891.—*Greyish black powder*, soluble in cold water to an olive-brown solution ; in alcohol to an olive-brown solution having a faint, greenish tint ; in acetic acid to an olive-green solution ; in strong sulphuric acid to a dark crimson solution, on diluting which with water a green precipitate falls out. Hydrochloric acid added to the aqueous solution throws down a dark olive - green precipitate, while caustic soda turns the colour of the solution to a dull red.—*Dyes* cotton and wool from salt baths olive-greens ; not altered by acids, turned brown by alkalies ; moderately fast to soaping and light.

Chrome Blue—

(BAYER.)—1891.—*Blue-black paste;* with water it gives a deep blue solution ; with alcohol and acetic acid an azure blue solution ; strong sulphuric acid forms a scarlet solution, which on dilution with water turns to a dark red turbid solution. Hydrochloric acid turns the colour of the aqueous solution to scarlet, while caustic soda changes it to a violet.—*Used* in calico printing, giving, with chrome acetate mordant, violet-blue shades, fast to soaping.— *Dyes* chrome mordanted wool violet-blue shades, turned red by acids ; fast to alkalies ; bleeds slightly to soaping.

Chrome Violet—

(BAYER.)—1891. —*Dark brown paste*, which on being heated with water turns greenish ; alcohol dissolves it to a deep violet solution, acetic acid to a dark green solution ; strong sulphuric acid forms a yellow-brown solution from which, on dilution with water, a dark red orange precipitate falls out. Hydrochloric acid changes the colour of the aqueous solution to a deep orange, while caustic soda throws down a red-violet precipitate.— *Used* in calico printing, giving with chrome acetate mordant reddish violet shades, fast to soaping.—*Dyes* wool mordanted with chrome red violet shades, turned red by acids ; fast to alkalies ; bleeds slightly when boiled in soap.

Cinereine—

(POIRRIER.) — 1890. — *Grey crystalline powder*, slightly soluble in cold water, easily in hot water to a dark violet solution ; in alcohol to a violet solution ; in acetic acid to a dark violet solution ; strong sulphuric acid dissolves it with a dark violet colour ; on dilution with water the colour changes to a blue-violet solution. Hydrochloric acid has no action on the aqueous solution, while caustic soda throws down a violet-red precipitate.— *Dyes* tannin mordanted cotton silver-grey to dark grey shades, turned dark blue by acids, reddish blue by alkalies ; fast to soaping and light. Can be used in calico printing as a tannic colour.

Congo Rubine—

(BAYER, ACTIENGESELLSCHAFT.) — 1891.
—*Dark brown-red powder*, soluble in water to a deep crimson solution ; in strong sulphuric acid to a blue solution, from which, on dilution with water, a blue precipitate falls down. Hydrochloric acid added to the aqueous solution throws down a blue precipitate ; caustic soda throws down a purplish blue precipitate.
—*Dyes* unmordanted cotton from a boiling salt bath, or soap and sulphate of soda bath, lilac red to magenta shades, turned blue by acids ; fast to alkalies and soaping ; not fast to light.

Direct Yellow—

(A. FISCHESSER.) — 1891. — *Ochreous-coloured powder*, soluble in water to a turbid brownish yellow solution ; slightly soluble in alcohol ; acetic acid changes the colour to a dark orange brown ; soluble in strong sulphuric acid to a brown solution, from which, on dilution with water, a brownish yellow precipitate falls out. Hydrochloric acid added to the aqueous solution throws down an orange-brown precipitate ; caustic soda reddens the colour of the aqueous solution.—*Dyes* unmordanted cotton from a soap and salt bath fine greenish yellow shades ; wool and silk from salt baths greenish yellow shades, turned buff by acids ; darker by alkalies ; not quite fast to soaping ; moderately fast to light.

Fast Acid Violet 10B—

(BAYER.)—1891.—*Dark violet powder*, soluble in water to a violet-blue solution ; in alcohol to a violet, and in acetic acid to a violet-blue solution ; strong sulphuric acid forms a deep yellowish brown solution, which on the addition of water turns first brown then green. Hydrochloric acid turns the colour of the aqueous solution first green then brownish yellow ; caustic soda produces no change.—*Dyes* wool and silk from acid baths violet-blue shades, turned first green then orange-yellow by acids ; bleeds slightly when boiled in soap and water. Can be used in printing wool or silk.

Fast Neutral Violet Extra—

(CASSELLA AND CO.)—Ethylated neutral violet. (1891.)--*Dark brown-violet liquid*, with a coppery lustre, soluble in water to a violet-red solution ; in alcohol to a red-violet solution ; in acetic acid to a violet-red solution, redder than the aqueous solution ; strong sulphuric acid gives a brown-violet solution, which on dilution with

water turns violet. Hydrochloric acid changes the colour of the aqueous solution to a bright reddish brown ; caustic soda has no action on the aqueous solution.—*Dyes* tannin mordanted cotton, wool and silk from neutral baths fine reddish violet, turned greenish blue by strong acids ; not altered by dilute acids ; reddened by caustic soda ; bleeds slightly when boiled in soap and water ; not quite fast to light.

Flavazol—

(ACTIENGESELLSCHAFT.) — 1891. — *Dull orange-yellow powder* ; soluble in water to a lemon-yellow solution ; in alcohol to a yellow solution ; slightly soluble in acetic acid to a pale yellow solution, the powder being turned buff colour. Strong sulphuric acid forms a deep Bismarck Brown solution from which water throws down a buff - coloured flocculent precipitate. Hydrochloric acid added to the aqueous solution throws down a faint reddish buff-coloured precipitate ; caustic soda gives a clear yellow solution.—*Dyes* chrome or alumina mordanted wool olive-yellow to yellow shades, slightly reddened by acids ; fast to alkalies and soaping.

Formyl Violet S, 4B—

(CASSELLA AND CO.) — 1891. — *Purple powder with a brownish lustre*, soluble in water to a bright violet solution ; in alcohol to a violet solution ; in acetic acid to a blue solution ; in strong sulphuric acid to an orange-yellow solution, which on diluting with water turns first yellow then, on standing, becomes colourless. Hydrochloric acid changes the colour of the aqueous solution to a pale greenish yellow, which becomes colourless on standing ; caustic soda changes the colour to a pale blue.—*Dyes* wool from acid baths bright blue shades of violet ; turned orange by strong acids, green by dilute acids, not altered by alkalies ; not quite fast to soap ; moderately fast to light. Can be used in calico and wool printing, but does not give fast shades.

Gallic Indigo S—

(DURAND AND HUGUENIN.)—A compound of gallocyanine and aniline. (1891.)— *Bronzy blue crystalline powder*, soluble in water to a blue solution ; in alcohol to a bright blue solution ; in acetic acid to a pale reddish violet solution ; in strong sulphuric acid to a deep reddish blue solution, turning violet-red on diluting with water. Hydrochloric acid added to the aqueous solution turns it to a deep reddish violet ; caustic soda throws down a blue precipitate.—*Dyes* wool and silk from acid baths fine indigo blue shades, fast to soaping ; not acted on by acids ; turned brown by alkalies. Fast to light.

Gambine Yellow—

(READ HOLLIDAY AND SONS.)—1890.— *Dull orange-red paste,* insoluble in cold but soluble in hot water to a yellow solution ; in alcohol to a reddish amber solution ; insoluble in acetic acid ; soluble in strong sulphuric acid, with a dark brownish yellow colour ; on dilution with water the colour becomes paler. Hydrochloric acid throws down from the aqueous solution a brown flocculent precipitate ; caustic soda forms an amber-coloured solution.—*Dyes* chrome mordanted wool olive shades of yellow, turned red by acids, purplish brown by alkalies ; fast to soaping and light. Can be used in calico and wool printing, with chrome or alumina mordants, giving moderately fast shades of olive-yellow, those with alumina being the brightest.

Gambine Yellow R—

(READ HOLLIDAY AND SONS.)—1891.— *Brownish orange paste;* the colouring matter is insoluble in cold water, but easily soluble in hot water to a brownish yellow solution ; in alcohol to a reddish amber solution ; insoluble in acetic acid ; soluble in strong sulphuric acid to a deep yellowish brown solution ; on diluting with water the original colouring matter is precipitated. Caustic soda forms an amber-coloured solution. — *Dyes* chrome mordanted wool and silk olive shades of yellow ; with alumina it gives brighter yellows ; fast to soaping and light ; turned scarlet by acids, orange by alkalies. Can be used in calico printing, with either alumina or chrome mordant, giving fairly fast shades of yellow.

Indoline R—

(GILLARD, P. MONNET, AND CARTIER.)— 1891.—*Black powder with a brownish tinge,* soluble in water to a purple solution ; soluble in alcohol and acetic acid to purple solutions ; soluble in strong sulphuric acid with a green colour which turns purple on diluting with water. Hydrochloric acid has no action on the aqueous solution, while caustic soda discharges the blue and leaves a faint brownish coloured solution.—*Dyes* wool and silk from acid baths indigo blue shades ; fast to acids ; turned brown by caustic soda ; fast to soaping ; moderately fast to light.

Methylene Green Ex. Yellow—

(MEISTER, LUCIUS.)—

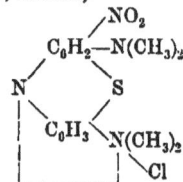

$$C_6H_2 \begin{matrix} -NO_2 \\ -N(CH_3)_2 \end{matrix}$$

$$\begin{matrix} N & & S \\ & C_6H_3 & \\ & & N(CH_3)_2 \\ & & Cl \end{matrix}$$

(1891.)—*Dark brown powder with a bronzy lustre,* soluble in water to a deep blue-green solution ; in alcohol and acetic acid to deep blue solutions ; strong sulphuric acid dissolves it to a pale green solution which is not altered on adding water. Hydrochloric acid added to the aqueous solution turns it yellower, caustic soda bluer.—*Dyes* tannin mordanted cotton, wool and silk from neutral baths dark yellow-green shades, turned yellower by acids, bluer by alkalies ; fast to soaping. Used in calico printing with a tannic mordant, giving fast shades.

Naphthyl Blue—

(KALLE.) — 1891. — *Purple, somewhat bronzy, powder,* soluble in water to a bright blue solution ; slightly soluble in alcohol to a deep blue solution; in acetic acid to violet-blue solution; in strong sulphuric acid to a deep green solution, from which, on dilution with water, a blue precipitate falls down. Hydrochloric acid added to the aqueous solution throws down a blue precipitate, while caustic soda throws down a violet precipitate.—*Dyes* wool from an acid bath reddish blue shades ; fast to acids, alkalies and soaping.

Naphthyl Violet—

(KALLE.) — 1891. — *Purplish black powder,* soluble in water to a dirty red violet solution ; in alcohol to a dark red violet solution ; in acetic acid to a red violet solution ; in strong sulphuric acid to a dark green solution, from which, on diluting with water, a dark violet precipitate falls down. Hydrochloric acid and caustic soda throw down red violet precipitate from an aqueous solution.— *Dyes* wool from acid baths red violet shades, fast to acids, alkalies and soaping.

Persian Red—

(A. FISCHESSER.)—1891.—*Bright scarlet powder,* slightly soluble in cold, easily in hot water to a deep scarlet solution ; insoluble in alcohol and acetic acid ; soluble in strong sulphuric acid to a deep crimson solution ; on diluting with water a dull scarlet precipitate falls out. Hydrochloric acid added to the aqueous solution throws down a scarlet precipitate ; caustic soda changes the colour of the solution to a dark scarlet.—*Dyes* wool and silk from acid baths, also on chrome mordanted wool fine Bordeaux Red shades, fast to acids, alkalies and soap ; moderately fast to light. Can also be used in calico printing with a chrome mordant.

Polychromine B—

(GEIGY.)—Prepared from polychromine, which is similar to Primuline in its composition by the action of ammonia. (1890.) *English Patent* No. 15,671, 1890.—*Brickred powder*, soluble in water to a reddish orange solution ; in alcohol to an amber-coloured solution ; in acetic acid to a faint reddish brown ; in strong sulphuric acid to a dark red solution which, on diluting with water, turns to a pale yellow turbid solution. Hydrochloric acid added to the aqueous solution throws down a dark purple precipitate, while caustic soda is without action.—*Dyes* unmordanted cotton from a salt bath orange shades of yellow, turned blackish purple by acids, brown by caustic soda ; fairly fast to soaping ; not fast to light. By passing through a bath of sodium nitrite the colour on the cloth is diazotised, then it can be developed into new shades ; thus with b-naphthol it gives a brown, fairly fast to soap and light.

Rosinduline SS—

(KALLE AND CO.) — 1891. — *Scarlet powder*, soluble in water to a crimson solution ; in alcohol it is only slightly soluble to a pale crimson solution ; in acetic acid to a crimson solution ; in strong sulphuric acid to a bright green solution, from which, on diluting with water, a pink, somewhat gelatinous, precipitate falls down. Hydrochloric acid added to the aqueous solution throws down a crimson precipitate, while caustic soda has no action.—*Dyes* wool and silk from acid baths bright crimson shades, fast to acids and alkalies ; bleeds slightly on soaping ; moderately fast to light.

Salmon Red—

(ACTIENGESELLSCHAFT.)—1891.—*Scarlet powder*, slightly soluble in cold, easily in hot water to scarlet solution ; in alcohol to a turbid pale scarlet solution. Acetic acid turns the powder purplish brown ; strong sulphuric acid gives a violet solution from which, on addition of water, a dull purple precipitate falls down; hydrochloric acid throws down a dull crimson precipitate from the aqueous

solution, while caustic soda throws down a faint red precipitate.—*Dyes* cotton, wool or silk from salt baths salmon to orange-brown shades, turned violet-red by dilute acids ; yellower by alkalies ; fast to soaping.

Thionine Blue—

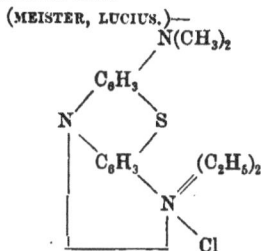

(MEISTER, LUCIUS.)—

$$N(CH_3)_2$$
$$\diagup$$
$$C_6H_3$$
$$\diagup \qquad \diagdown$$
$$N \qquad\qquad S$$
$$\diagdown \qquad \diagup$$
$$C_6H_3 \qquad (C_2H_5)_2$$
$$\diagdown \qquad \diagup$$
$$N$$
$$|$$
$$Cl$$

1891.—*Greenish bronze powder*, soluble in water to an azure blue solution ; in alcohol and acetic acid to azure blue solutions with a greenish reflex ; strong sulphuric acid gives a deep yellow-green solution which, on diluting with water, turns a deep blue ; hydrochloric acid turns the colour of the aqueous solution greener, while caustic soda turns it redder.—*Dyes* tannin mordanted cotton, wool and silk from neutral baths bright blue shades, turned greenish by acids, reddish by alkalies ; fast to soaping.

Titan Blue—

(READ HOLLIDAY AND SONS.)—1891.—*Black-blue powder*, readily soluble in water to a violet-blue solution ; in alcohol to a violet-red solution ; in acetic acid to a violet, and in strong sulphuric acid to a bright blue solution, from which, on addition of water, a blue precipitate falls down. Hydrochloric acid throws down a blue precipitate from the aqueous solution, while caustic soda turns the colour of the solution to a bright scarlet.—*Dyes* unmordanted cotton from a salt bath bright shades of violet-blue, turned dark blue by acids, bright red by caustic soda ; not fast to soaping.

Toluidine Blue—

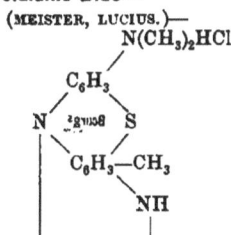

(MEISTER, LUCIUS.)—

$$N(CH_3)_2HCl$$
$$\diagup$$
$$C_6H_3$$
$$\diagup \qquad \diagdown$$
$$N \qquad\qquad S$$
$$\diagdown \qquad \diagup$$
$$C_6H_3{-}CH_3$$
$$\diagdown$$
$$NH$$
$$|$$

1891.—*Yellowish bronzy powder*, soluble

in water to an azure blue solution ; in alcohol to a similar solution but having a greenish reflex ; in acetic acid to a purplish blue ; in strong sulphuric acid to a deep yellow-green solution which, on diluting with water, turns blue ; on adding hydrochloric acid to the aqueous solution the colour becomes bluer, while an excess of acid turns it a green-blue. Caustic soda throws down from the aqueous solution a brown-violet precipitate.—*Dyes* tannin mordanted cotton, wool and silk from neutral baths bright shades of blue, turned green by acids, red by alkalies, which are not fast to soaping.

Vacanceine Blue—

(READ HOLLIDAY AND SONS.)—1891.— *Blue-black liquid*, very slightly soluble in cold water ; more soluble in hot water to a dark red-violet solution ; in alcohol and acetic acid to dark red-violet solutions ; in strong sulphuric acid to a dark green colour ; on diluting with water a very dark blue-black precipitate falls down. Hydrochloric acid added to the aqueous solution throws down a blue-black precipitate, while caustic soda throws down a reddish black precipitate.—*Dyes* tannin mordanted cotton indigo-blue shades, turned dark blue by acids, dark reddish blue by alkalies ; fast to soaping.

Victoria Black G—

(BAYER.)—1891.—*Reddish black powder*, soluble in water to a very dark, dull crimson solution ; in alcohol to a turbid blue solution ; in acetic acid to a dull, dark purple solution ; in strong sulphuric acid to a dark green solution, from which, on dilution with water, a dark red precipitate falls out. Hydrochloric acid added to the aqueous solution throws down a dull crimson precipitate, while caustic soda throws down a purple precipitate. — *Dyes* wool and silk from acid baths blue-black shades, fast to acids, alkalies, soaping and light.

Victoria Black-Blue—

(BAYER.) — 1891. — *Dark purple-black powder*, soluble in water to a dark, almost black, purple solution ; slightly soluble in alcohol to a dark purple solution ; slightly soluble in acetic acid to a dark violet-black solution ; soluble in strong sulphuric acid to a dark green solution, from which, on diluting with water, a blackish blue precipitate falls out ; the same precipitate is obtained by adding hydrochloric acid to an aqueous solution. Caustic soda throws down from the aqueous solution a dark purple precipitate.—*Dyes* wool and silk from acid baths black-blue to blue-black shades, fast to acids, alkalies, soaping and light.

www.ingramcontent.com/pod-product-compliance
Lightning Source LLC
Chambersburg PA
CBHW021823190326
41518CB00007B/725